HISTORY OF BROADCASTING: RADIO TO TELEVISION

HISTORY OF BROADCASTING: Radio to Television

ADVISORY EDITOR

Dr. Christopher Sterling, Temple University

EDITORIAL BOARD

Dr. Marvin R. Bensman, Memphis State University
Dr. Joseph Berman, University of Kentucky
Dr. John M. Kittross, Temple University

Television

A Struggle for Power

FRANK C. WALDROP AND JOSEPH BORKIN

ARNO PRESS and THE NEW YORK TIMES
New York • 1971

Reprint Edition 1971 by Arno Press Inc.

Reprinted from a copy in The Newark Public Library

LC# 72-161140
ISBN 0-405-03561-6

HISTORY OF BROADCASTING: RADIO TO TELEVISION
ISBN for complete set: 0-405-03555-1
See last pages of this volume for titles.

Manufactured in the United States of America

TELEVISION
A *Struggle for Power*

TELEVISION

A STRUGGLE FOR POWER

by
FRANK C. WALDROP

and

JOSEPH BORKIN

with an introduction by
GEORGE HENRY PAYNE
Member of the Federal Communications Commission

New York · 1938
WILLIAM MORROW AND COMPANY

TELEVISION
COPYRIGHT - - - - - - 1938
BY FRANK C. WALDROP AND JOSEPH BORKIN

All rights reserved. This book, or parts thereof, must not be reproduced without permission of the publisher.

PRINTED IN THE UNITED STATES
BY THE STRATFORD PRESS, INC., NEW YORK

Introduction

THE MANUSCRIPT OF THIS BOOK WAS SHOWN TO ME AFTER IT had been virtually completed, and I was then, as I am now, enthusiastic about the work. There are, of course, many statements that I would not have made myself, but there is room for many opinions and theories on so living a subject as radio. The very fact that the subject is so controversial indicates that it is bristling with life. Free discussion of ideas should, of course, be encouraged for it is from such discussions that democracy takes its nourishment.

The present work is one of the few which have been written from the point of view of the public to whom the ether actually belongs; but that is only one of its virtues. It is full of information that has never been in print before; it has a fresh point of view and the treatment of the material has pleasant ease, clarity and finish. The authors present not only a wealth of valuable information, but also show intelligence, talent and vision, and what is even rarer, courage.

After three years on the Federal Communications Commission I had begun to conclude that the public was so inured to the evils of broadcasting that it would be a long time before it would awake to protect its own interests. Suddenly the picture changed. In the last six months many things have happened to show that there was not the apathy that many of us in Washington who are interested in reform, had been made to believe existed. I have seen this interest growing and now I see that it is beginning to express itself with no little clarity and some indignation.

In TELEVISION: *A Struggle for Power*, the authors are warning the public of the future. They deal not with the demoralization of the home, which is to me more important than the question of monopoly, important though that question be. Important, however, is the fact that they have written lucidly and suavely an unheard-of story. To at least one member of the F. C. C. most of it was news.

It was Professor Doriot at Harvard University, who, after my first lecture there three years ago, asked me to consider what was apt to be the effect emotionally and mentally on a nation when the same message would be delivered in thirty million homes simultaneously and that message was one calculated to arouse great antipathy or hatred of some given object, person, race, nation or religion. His point was that when there was no opportunity for counter-argument or counter-propaganda there was something dangerous in the fact that practically an entire nation might be stirred to its emotional depths before there was an opportunity for counter-argument, contradiction or reflection.

Even with the many engrossing duties connected with communications, this grave question began to formulate itself in my mind, which formulation resulted in a paper on "The Home versus the Radio" and an address at the Chicago Educational Conference, responding to the question, "What Shall We Do with Radio?"

INTRODUCTION vii

We have heard the phrase "public interest, convenience, or necessity" bandied about in and out of season. It has been pulled by the ears and pushed by the scruff of the neck into almost every hearing, oral argument, and everything else relating to radio licenses that has come up before the Communications Commission. The phrase evidently was intended originally to furnish armor plate for the protection of the public's interest in its last invaluable asset, the radio frequency.

The little significance that this collocation of words has come to have in the practical affairs of radio is astonishing. At the innumerable hearings I have sat through and in the innumerable reports on radio matters I have read, the expression has never amounted to anything more than a canting phrase, which has long ago lost its original vitality and significance, if it ever had any. The questions actually asked at such hearings are "Will the station be able to make money?" "Will the station be able to get advertising?" The question whether the station will contribute to the culture of the people, to their mental life or their welfare otherwise, seems to occur to no one.

What, therefore, has become of the significance of this expression which is so much used in our Commission? Its meaning, in its degenerate state, is commercial in essence, referring to the interests of those who hold Government concessions, and not to the interests of the public. By what mental gymnastics the phrase was made to do a complete somersault, it is difficult to say. Some easy-going officials have doubtless contributed their share during the past ten years to the distortion of this protective phrase of the people. Persistent and astute attorneys, fighting for their clients' interests and watching those interests with the thousand eyes of an Argus, have contributed even more.

Unfortunately, there are no organized powers to fight against the invasions of the people's rights from day to day as they occur. Such things are done stealthily and as a rule by small aggressions and small changes, until the public awakes to find itself separated from one more of its possessions.

So entrenched have the commercial interests become in radio that a great deal of strong affirmative legislation might be necessary to protect the public, which has been so persistently stripped of its rights, against further encroachments. It may be necessary, therefore, for Congress to restate with a finer definition the rights of the people in the radio frequencies, so that no officials or attorneys can distort the meaning. The phrase, in the first place, is a nebulous collection of words. The word "welfare," in describing the public interest, would, I believe, be less susceptible to misinterpretation and distortion.

Perhaps this book will help you to determine what measures you, the citizen, want taken. This is an interesting book and an important one. It tells an amazing story, at times a sordid one, with brilliance. Whether I, if these researches had been my own, would have reached the same conclusions, I cannot say. But no one can deny that the authors have ability, integrity and a sincere desire to perform a public service.

GEORGE HENRY PAYNE

Washington, D. C.
February 7, 1938.

CONTENTS

		PAGE
	Introduction	v
	Preface	xi

CHAPTER		
I.	Prelude to Struggle	3
II.	In the Arena	11
III.	Inventing a Necessity	22
IV.	Wires versus Wireless	37
V.	New Public Property	44
VI.	The Inadequacy of Law	49
VII.	The Philosophy of the Spectrum	62
VIII.	Trouble in Heaven	70
IX.	The Ethereal Klondike	81
X.	Microphones and Censors	97
XI.	Ethics and the Listener	111
XII.	The Somnolent Cinema	121
XIII.	No. 195 Broadway	131
XIV.	The Bell System	139
XV.	The Belle of Hollywood	149
XVI.	RCA Pays a Dividend	162
XVII.	The Trust Dissolved?	177

XVIII.	Patents and Power	199
XIX.	Past Is Prologue	216
XX.	Return of a Pioneer	231
XXI.	The Seven Wise Men	241
XXII.	Public Policy	254
	Appendices	271
	Bibliography	277
	Notes	283
	Index	293

Preface

A PROPER REPORT ON THE STATE OF TELEVISION TODAY CAN be rendered only in terms of dynamics and change. For television comes upon the world not isolated but influencing and influenced by technical research; by the economics of telegraphy, telephony, newspapers, the stage, sound motion pictures and sound radio; by laws of Congress, acts of regulatory commissions and decisions of high courts; by programs of education and entertainment, free speech, censorship, private and public morals; and by the rights of individuals seeking reward for skill and genius at invention.

It is the object of this book to show, in the simplest possible manner, the relationships between those apparently diverse interests. In matters of technological change, the discussion of mere techniques has too long (and too often) been allowed to obscure the bearing of such changes upon the "public interest, convenience, or necessity." Too long have the masses of people accepted with placidity every purported change in scientific matters, thereby neglecting to exercise positive concern as to the possible effects upon themselves.

Radio communication is growing in power and perfection. In time it probably will become the chief means of conveying information to masses of people throughout the world. The authors can hope to contribute only some small part toward the accumulation of evidence which will convince those people that it is necessary not to allow control of this means to be exercised without their knowledge of how the change came about. In order to make this contribution with no sacrifice of accuracy or detailed record, however, it has seemed best to summarize in simple outline, early in the discussion, the historical background out of which television, both as science and as art, has grown. It is the authors' belief that their short digest of the principles by which television works will clarify not only the main issues of the text but the reader's understanding of the day by day struggle for power in the commercial spectrum.

Any merit this work may have is due in no small part to critical readings of the manuscript by Professor Myron W. Watkins, head of the economics department of University College, New York University; to the suggestions of Dr. H. C. Engelbrecht, author of *Revolt Against War*; and to the diligence of Pauline Marks, Eleanor Lanier, and Dorothy Sugarman in research and analysis of library material.

F. C. W.
J. B.

1. Prelude to Struggle

SOME PEOPLE LIKE TO SAY THAT TELEVISION IS THE CHILD OF art and science—and so it is, in a way. But, however blissfully art and science may be wedded, the child is no ordinary child, and government and money are deeply disturbed about its future.

To be exact, television represents a synthesis of scientific achievements by means of which electrical analyses of sounds and of the appearance of objects are blended and transmitted in a split second throughout wide areas. Television is just a trick, really; the trick of using electrons in order to look at something not visible to the naked eye. But through the perfecting of this trick the means of access to public credulity, and to the power which that access gives, lie open to some man's grasp—and not enough people know it.

Consider for a moment the report which a group of distinguished Americans, in all seriousness, lately gave to their government upon this matter:

When to the spoken word is added the living image, the effect is to magnify the potential dangers of a machine which can subtly instill ideas, strong beliefs, profound disgust and affections.

There is danger from propaganda entering the schools,

and perhaps much greater danger from the propaganda entering the home. How great is the power in the control of mass communication, especially when helped by modern inventions, has been made clear recently in countries that have had social revolutions, and which have promptly, in a very short period, brought extraordinary changes in the expressed beliefs and actions of vast populations.

These have been led to accept whole ideologies contrary to their former beliefs, and to accept as the new gospel what many outsiders would think ridiculous. The most powerful means of communication, especially for rapid action in case of revolution, are the electric forms like radio and television, which spread most skillfully presented ideas to every corner of the land with the speed of light and a minimum of propaganda labor. Compared with these, the impromptu soap-box orator with his audience of a dozen, or a local preacher with his 200, are at a grave disadvantage. Certainly no advertiser would expect to sell as many goods by amateurish appeal reaching 10 dozen as by a captivating one reaching 10 million.

Television will have the power of mobilizing the best of writers and scene designers, the most winning of actors, the most attractive of actresses.[1]

These soothsayers gave their views to the President of the United States on June 18, 1937, and signed their letter of transmittal, respectfully:

Harold L. Ickes, Harry H. Woodring, Henry A. Wallace, Daniel C. Roper, Frances Perkins, Harry L. Hopkins, Frederic A. Delano, Charles E. Merriam, Henry S. Dennison, and Beardsley Ruml. Even the most casual student of current affairs will recognize among them some astute analysts of ways and means to mold the public mind. As to the scientific import of their study, they told the President they drew upon the National Academy of Sciences, the Social Science Research Council, and the American Council on Education for expert testimony.*

Basic questions of national policy arise. We must know who shall and who ought to control television; what ideas and whose it shall convey. What will be the effect upon human institutions? Who should be rewarded or punished for having brought it upon us?

Perhaps there are no clear-cut answers to these questions. Perhaps there should not be. But at least there are some elements of fact to be had, some definitions of issues, which the citizen can employ as he wishes. The authors of this book propose no answer. But we can hope, and we do hope, that television may not be allowed to fall unknowingly into the hands of some plotmaker, some group in power or seeking power to destroy democracy in this time when man, fretted by his inventions, cannot bear either to throw them away or put them to use according to a conscious plan.

This is an important matter for the common man as well as the special pleader. Communication holds together the very fabric of society, and as social groups grow in power and complexity, their systems of exchanging information become infinitely more important and widespread. Perhaps the most recent major demonstration of the importance of communications to national interest was at Versailles, when the masters of destiny haggled for three things above all others: oil, international communications, and its twin, international transportation. The nations of today are ranked

* The National Resources Committee, of which these signatories are members, has made important studies not only of technological trends, but of water power, land uses, and other basic instruments for development of better living conditions.

in power according to their standing in those three categories.

Television happens to be the newest, and at the same time the most effective, means of communicating information, misinformation, and entertainment. To attempt a summary report of its technical status is dangerous, for scientific standards and conceptions are changing rapidly. Hence we tell here only what was disclosed or reliably reported as of January, 1938.

The approximate standard performance offered sharp, clear pictures upon a glass screen seven inches high by twelve inches wide. Experimenters had succeeded in projecting enlarged reproductions upon screens as great as three by four feet in area for home use; and some demonstrations had been given on screens nine by twelve feet, and were improving steadily.[2]

Reproductions were still in shades of black and white, insofar as routine broadcasts were concerned, but color television was already a practical accomplishment in public display in England and the engineers were busy with even stranger things. Men were actually searching for the mathematical outlines of ways and means to transmit sensations of smell and feeling by electricity.[3] Quite soberly, the question was put in the report to the President whether we may not some day sit in our libraries and have presented to us the electrical reproduction of distant scenes, in natural colors and sounds, and with every aspect of smell and feeling except actual substance.

Small wonder, then, that his cabinet officers and their technical advisers suggest to the President of the United

States that the people of the country should begin to think of what they would like to do with television. It is far more than a propaganda device for the plausible orator. True, it is invaluable to him. Imagine the candidate for public office standing full length in your living room, pointing to charts, beaming his smile, and exuding the fragrance of roses or stale cigars. Imagine a tottering rule of the elders being restored by some father of his country, long dead but able through electrical reproduction of his living manner to adjure his countrymen once again to avoid evil doctrines and to stray not on strange ways. This is no mad notion. Your phonograph record is a prison for sound. The motion picture film is an imprisonment of sight and sound in one. Grant the engineers their capture and transmission of smell and feeling by electricity, and who will say they can never imprison those sensations, too, for enduring record and reproduction?

But we need not wait until the problems of communicating smell and feeling have been solved to feel the force of this new instrument upon our lives. Sound radio has already indicated the way in which changing technical methods of communicating information affect existing human institutions. Radio, it is by now generally recognized, is a rival of the daily press both for the privilege of distributing news and for revenue from advertisers. In 1937, as the radio industry continued the rapid expansion begun about 1923, newspaper industrialism continued a decline started at the same time.

The trend in radio is indicated by the 1937 financial history of the National Broadcasting Company, the largest sin-

8   TELEVISION

gle program service organization. In that year NBC showed a revenue gain of eighteen per cent over 1936, and with two other program companies hired one thousand additional musicians. Thirty-six new broadcasting stations came into operation.[4] In contrast, while the newspaper industry, with a daily publication of 41,400,000 copies, took in $595,000,000 of revenue, it showed a bare two per cent gain over 1936. There were 2084 daily publications in English in 1937, a decline of twenty-three from the previous year; 10,629 weekly journals, a decline of 176; and 359 semi-weeklies, representing a loss of eighteen.[5]

But journals and publications have not yet felt the worst of competition from radio. One of the most important by-products of television is the "facsimile recorder," an instrument which will print messages of record in response to electronic impulses. It can be operated by business establishments to replace telephone and telegraph leased wires between branch establishments, and also to print news in the home. The television facsimile machine responds to a radio signal in the same basic manner that the sound radio receiving set now does. Indeed, it is designed to be attached thereto. The electrical impulse causes a stylus to sweep across plain white paper and bring out not only script or printed letters but also reproductions of photographs, both in black and white and in color combinations. Facsimile machines, operated in conjunction with central broadcasting stations, are today literally capable of producing the newspaper in the home, eliminating two of the greatest expenses now attached to the publications industry, printing and delivery. The effect such a radical alteration in methods must have upon investments in presses, trucks, and buildings is obvious. The effect upon employment is equally apparent. Knowledge of these facts may make for understanding of why newspaper publishers are so eager for radio stations, today, and already hold approximately one fourth of all the licenses of operation granted by the Federal Government. They are simply trying to shift their fortunes with the tide of technology.

As in publishing, so in many other trades, industries, and affairs of men. Television already has advanced to such a state of perfection that it can be, and in some countries is, used to amuse. It can present a play in your home just as the drama proceeds in a studio or for that matter in an ordinary theatre. It can report events as they happen. It can fill in the time between by reproduction of motion pictures already caught on film. Mariners have found ways to use certain modifications of television to overcome fog. Soldiers and sailors use it to spot gunfire, and scenes are transmitted from ship to shore, from plane to ground, with clarity and regularity.

Obviously, a device of such powers is not going to be allowed to fall into the hands of one group or another uncontested. Television is an instrument not only of great potential power and uses, but of profit.

Some of the greatest corporate organizations in the modern world are preparing, indeed even now are fighting, to control its development. The American Telephone and Telegraph Company, the Radio Corporation of America, Westinghouse Electric and Manufacturing Company, General Electric Company, Columbia Broadcasting System—these are aristocrats in our financial oligarchy. And they are well aware that if any one of them is allowed to control the growth of television, extinction is threatened for the others.

Inventors are searching passionately for solutions to last details in support of patent claims. Zworykin, Baird, Farnsworth, Finch, Lubcke, Round, Alexanderson, Armstrong, De Forest, LaMert—men whose names for the most part mean nothing to the general public—are the makers of the future for that public.

Not all who are deeply involved in these developments realize what is happening to them. The great motion picture industry and its dependencies, such as the thousands of picture exhibitors, are relatively passive in the face of change. Western Union, Postal Telegraph, and Mackay Radio, in contrast with the great Bell telephone system, seem unable to organize themselves for adequate defense.

But whether they like it or not, television presses change upon them, every one.

2. In the Arena

IT IS ONLY NATURAL THAT THE OPERATORS OF THE PRESENT sound radio system have assumed that television will gravitate into their hands. After all, they argue, their money has financed the present development. In some degree they are right. Furthermore, they are the financial underwriters of most of the research which eventually must result in obsolescence of the very plant and structure which earns profits by means of which to endow research. This contention is less accurate. Corporation engineers appear to have contributed very little to the fundamental development of television. But of one thing there can be no doubt. Unless business men are permitted to swing the basis of financial development along with the change in technical means of operation, chaos must certainly ensue.

Radio is no longer an infant industry. Electrical communications operate within the framework of complex and important corporate organizations. Great fortunes depend upon right judgment and delicate maneuvering, and maneuvering now, for the status of television today is such that unless the operators of radio, the movies, and several other industries show considerable speed and intelligence, they may find their corporate horses shot out from under them. Television has a popular appeal.

Its progress in England is most commonly referred to, generally because in that nation the government has acted to force operations in which the public can take a part. Widely publicized demonstrations of the reporting of events as they occur, such as the showing of the final tennis matches for the Davis Cup, and the Armistice Day exercises of 1937, caused sensations in the United States where television is not operated so openly for the public. The British programs are sent out from Alexandra Palace, London, and are generally made up of motion picture films which have been exhibited in theaters at least three months prior to the time of televising; of vaudeville skits and dramatic presentations; of "radio visits" to scenes of historic interest and beauty; and of spot news occurrences, such as the Coronation and other State functions. It is highly significant that public interest in British television became intense only after the showing of actual news incidents was possible.

At the beginning of 1937 less than one thousand receivers were in operation, but in December of that year the BBC reported nine thousand licensed receiving sets in daily operation within service range of Alexandra Palace.[1] During the annual radio show, "Radiolympia," sets went on sale at between $178 and $200. The BBC system has been developed on the basis of a government commission's recommendations made in 1935. At that time all inventors and engineers were ordered to place their patents in a common pool from which each could draw the patents of all the others on a royalty basis to build sets and equipment according to his notion of what would be the best.[2] To insure a minimum of good operation, the BBC fixed standards of performance for both broadcasting and reception, with the cost of operation met out of the revenues from taxation of sound radio sets.

BBC has set a standard of performance in accord with the recommendations of 1935, as follows: Programs * are broadcast with a peak power of 17 kilowatts on a 45 megacycle channel for the visual program, and 3 kilowatts on a 41.5 megacycle channel for accompanying sound. The official service range is encompassed within a radius of thirty miles from Alexandra Palace but good reception is reported as far away as Ipswich, seventy miles from the tower. Coaxial cables are being laid from London to Birmingham and other cities to provide provincial service. The pictures are shown at the rate of fifty frames a second, of four hundred and five lines each—very sharp and clear definition.[3]

A public demonstration at the Dominion Theatre in London last January drew an audience of three thousand. Pictures were projected on a screen six by eight feet.[4]

Television received little public attention in France until last year when, suddenly, the Ministry of Posts, Telegraph and Telephone announced that it had ordered the world's most powerful (30 kilowatts) transmitter to be erected in the Eiffel Tower, eleven hundred feet above the earth.[5] British Broadcasting Corporation officials were considerably disturbed, for the French government said it would permit commercial programs, and these, coming from such a powerful transmitter, might easily interfere with the BBC network.

* The reader who is not familiar with technical aspects of electrical communications need not feel concerned. In general, it might be said that when two systems are being compared, the one which is described in terms of larger numbers is the superior. A general analysis of television operation is given within the next few chapters.

The New York Times gave a clue to the conflict within, in a dispatch stating that the French announcement had "stirred speculation in American radio circles whether or not this move augured an international television race comparable to the one now being run in super-power broadcasting." [6] This conflict was adumbrated as early as 1933 when a British program was broadcast to a theater crowd at Copenhagen.[7] Undoubtedly, as technical proficiency advances some compromises will have to be made on television frequencies for international broadcasts along the lines now existing in sound radio. The French station is four times as powerful as any now licensed in the United States. No data are available in regard to picture definition.

In Germany television is being operated very efficiently by the Post Office Department. Exactly how many transmitters are in use is not known, but at least one is in Berlin, broadcasting on a radius of sixty kilometers of service range. Another in the Harz mountains has a radius of one hundred and twenty kilometers.[8] Five companies are manufacturing sets with large screen, cathode ray projectors, with picture definition of one hundred and eighty lines, twenty-five frames per second.[9] Very successful work was done during the Olympic Games in catching action scenes. One of the most interesting German developments is that of public television-telephone service, by means of which a person may see the one to whom he is talking by wire. For some time such a service has been maintained between Berlin and Leipzig, and last year the government authorized extension to several other cities.[10]

Russia is reported to have erected television transmitters of low caliber definition in Moscow, Kiev, and Leningrad, but apparently intends to go in heavily for further development, probably for military purposes. An order was placed with the Radio Corporation of America in 1937 for a 7.5 kilowatt station costing one million dollars.[11] An unofficial report stated that RCA was also retained to make receiving apparatus and that contracts were made for the use of RCA patents.[12]

In Italy, SAFAR, the authorized manufacturer of television equipment, reports technical efficiency on a standard performance of twenty frames of three hundred and seventy-five lines each per second. A chain of stations connected by coaxial cables and operating on service ranges of twenty-five miles radius each is reportedly being considered by the government.[13]

As early as 1932, the Japanese Radio Broadcasting Association claimed to be able to exhibit pictures on screens eight by twelve feet, but of undisclosed definition. Stations were being maintained by the government at Waseda and Hammatsu Universities, but so far as can be learned they were of narrow range and low definition.[14] Since then, considerable progress appears to have been made. It is expected that by 1940, when the Olympic Games are scheduled to come to Tokyo, a public broadcasting system will be in operation covering a twelve mile service radius.[15]

In Poland, Czechoslovakia, Holland, and Sweden television experimentation was progressing last year. Each of these countries has a transmitter offering experimental programs to the public, but no original research or basic patentable discoveries were reported.[16]

The important characteristic of television abroad is that in every country it is being conducted by government departments or in close co-operation with the government. In most cases, direct governmental subsidy underwrites the laboratory work and the public operations as well.

In the United States television is unquestionably more advanced, with respect to station operations and technical efficiency, than anywhere else in the world. There are eighteen licensed stations; and because so little is known about them by the general public, we have listed them in detail.*

What are the powers of television? It permits the leaving of messages. It is powerful in scattering persuasive arguments among masses of people. And the same electronic means that produce television permit multiple telephone conversations by radio or wire. And this is not all. Television informs, entertains, spies for gunners, guides mariners, prints newspapers. And not only the publishing industry may anticipate corporate corrosion as a result of its workings. Consider the case of the amusement trades. The clown never got rich from one performance. He collected his pennies in weary travel from village to village. The stars of Broadway grew proud as the lines grew long in front of box offices, for they knew that long lines meant long runs, long lives for plays and work for actors.

The movies changed all that. The clown's humble repertoire made just one short reel of laughter and was gone forever. He traveled no more, but let the can of film do the trouping. Broadway's stars and what has happened to them start no tears today. Everybody knows how Hollywood has ruined one Broadway and set up a thousand others across the nation until now we have a Broadway wherever there is a marquee and a billposter proclaiming next week's drama.

* See Appendix A.

But not everybody knows what is in the making for Hollywood. Suppose the clown's short and simple annal of amusement is presented to the whole nation in one brief moment. Its travels are over, once the nation's television sets have flashed the antic to all of America's homes in fifteen minutes flat. And travel is ended, too, for the great feature film. Why struggle downtown through traffic, then stand in line, and pay money to see *Mutiny on the Bounty*, when it can be enjoyed at home just as well?

The unhappy newspaper publisher, too, finds that by installing facsimile printers in people's homes to escape the expense of operating printing presses and delivery systems he only adds other burdens elsewhere. He cannot junk his machinery, turn the workmen out into the street, and go singing on his way. People demand support, whether or not they labor. In the United States, at least, a press free of censorship is guaranteed by the Constitution. But can the facsimile be called an instrument of the free press? That is an issue not yet settled for the publisher and he cannot face it with any certainty of success, for the Government holds firm control over facsimile's common carriers, as well as over those of all other electronic devices of communication. Moreover, the Government is aided in keeping its grip by the confused state of all thought concerning relationships between technology and human institutions. Not even in law can any basic definitions of private rights involving electric communications be given with certainty.

"Notions of sovereignty, states' rights, property, laissez faire, developed by land and commercial economics, are belied by the scientific facts of this novel method of communication," says the *Encyclopedia of Social Sciences*. "The peculiar characteristics of radio have evoked a distinct radio law, but the legal controls have been shaped largely by the state of the art, and require continual revision if they are to keep pace with its progress."

And we shall see, presently, that law has never been able to keep pace with art. Neither has government control of propaganda. One characteristic of the radio technology remains: if a program is broadcast, there is no way of being sure the wrong people will not hear it. This matter must be solved before television is as common as sound radio, or the absolutists are lost.

The untrammeled electron is at once a pleasure and a pain to the politician in power. By means of it he can address a whole people, but by that same means a whole people can be reached by his competitor if that competitor can gain access to the transmitter. In countries where absolutism is supposed to be the order of the day, the dictator commands his victims to attend the radio as faithfully as the Moslem heeds the muezzin's call to prayer. But he is tortured by the knowledge that some scoundrel from beyond the border, or even within it, is likely to commit piracy upon the sanctified domain and reach startled ears with unsanctified information.

To maintain their grip upon mass sentiment, indeed to forbid the exercise of intelligence, the iron men have been driven to some extraordinary measures. Japanese police, for instance, are charged with the responsibility of eradicating "dangerous thoughts." In Germany revolutionary suggestions from outside have become so common that the government is reported as planning a most extraordinary attempt to "save" the people by taking radio, as we know it, out of use. Early in August, 1937, there was an exposition in Berlin of high frequency radio developments, and that was very important. It revealed the purpose of changing the entire system of radio in Germany from wireless transmission to transmission by cables. The value of such an arrangement for war purposes is obvious; it would be safer from interruptions and against destruction. Besides, it is easier to control the programs of listeners, and to prevent the reception of "subversive" programs from other countries [17]—or even one's own. In March, 1937, programs being broadcast in Germany could be heard plainly in New York and Pittsburgh. Each period opened with a singing of the "Internationale," amounted to a harangue for development of a united Socialist-Communist front, and closed with the hymn of revolution. Hitler's agents combed the Fatherland, but if they ever found the daring broadcasters the world has not been told. The station happened to be mounted on wheels.[18]

Hitler's resolution to abandon radio broadcasting in favor of wires for both sound programs and television touches upon a basic problem of the whole industry, that of monopoly. But before we consider it, let's worry some more with the administrators of government. The cross-fire propaganda between warring dictators in Europe is a common topic of political conversation. Only lately have the gossips become aware that the cross-fire is no longer confined to one continent. Approximately forty programs are being broadcast from Europe daily in foreign languages and in English translations, intended exclusively for listeners in the Western Hemisphere. The Congressional Record offers as exhibits the mailing lists of American branches of the great salvation systems. To the names on these lists are sent cards every week advising the comrades at what hour and upon what radio frequency to heed the words of wisdom from afar. Toward the close of the year, broadcasts from foreign countries were arriving steadily not only from Europe but from Asia, Australia, and, of course, South America.

Nor was the Government of the United States allowing these extra-national campaigns to proceed unchallenged. For domestic consumption, the Department of the Interior's Bureau of Education organized a series of programs characterized by an opening hymn entitled, "Let Freedom Ring." Officials of the administration were found to rush before microphones at the slightest opportunity to explain every minor matter of policy, and the President of the United States was considered a professional master of the art of talking to a nation from the fireside.

For international consumption several powerful, privately financed stations were in operation. Their programs, however, were offered only after approval by the United States Government and were directed chiefly toward Latin America, in conformity with the "good neighbor" policy of the administration. Terms of highest praise for democracy, liberty, and other techniques of libertarian government considered disreputable elsewhere in the world were common characteristics of these broadcasts. The policy of allowing private interests to distribute these programs was distinctly declining in State Department favor by the fall of 1937. One bill pending in Congress proposed authority for the erection of a tremendously powerful government owned station directly dedicated to broadcasting of pro-American propaganda to the world.[19] Government departments clamored for equipment and broadcasting powers for propaganda, message exchange service, and for entertainment, pure and simple.

The question now is whether trends set up in sound radio will not prevail with television. The great technical issue of today is clearly indicated in the reference to plans for putting all German communication back into wires. Hitler's objective, naturally, would be to exclude interference of the sort that comes in pure radio operations. Every feature of television can be offered by wire distribution, except, of course, when it is used in connection with vehicles in motion. Shall communication in America be by wire or wireless? It is not entirely a simple matter of engineering technique. If television is confined to wired services, it is likely to continue expensive and therefore difficult to put into common service. If it is broadcast after the sound radio principle, then great areas of the nation may never receive it. Furthermore, a tremendous enterprise, that of radio operations in general, must be revised.

The Federal Government is being burdened with the multiple task of setting standards of performance, deciding between contestants for the right to perform, enacting legislation which will preserve all equities, and repelling political boarders, as it were, who seek to use sound radio and television to contaminate our institutions. Before we can guess what the government can do about it all, we must understand something of television's scientific structure.

3. Inventing a Necessity

THOUGH IT IS JUST NOW COMING INTO COMMON USE, TELEvision is far from a recent discovery. Experimentation in the combination of light-sensitive materials and electrical force were made as early as 1873, and in 1884 a German by the name of Nipkow laid down a principle of television operation which is the basis of all except the most recently developed types of machines.

There is a classroom truism to the effect that nobody knows what electricity is—which, unfortunately, tends to stifle ordinary discussion of any manifestation of electrical force. We may not know what electricity is, and we may be puzzled by its strange abilities, but we do, however, know some things about it—we know it is akin to light, to magnetism, and perhaps to heat.

The point to remember about radio communication is simply that it is a means of propelling a sound wave, an audio frequency, ever so much farther than it can go of its own accord, and of using a video—a sight—frequency to carry the appearance of an object to points beyond the powers of the unaided eye and light wave. The electromagnetic disturbance which does this work is known as the "carrier" frequency of a broadcasting station, and the whole trick of radio is in modulating the original sight or sound

INVENTING A NECESSITY 23

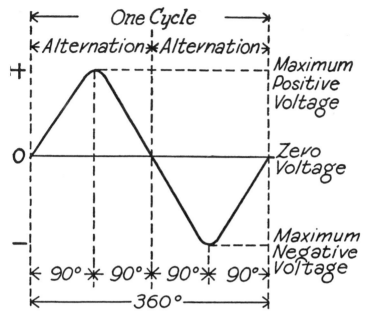

This is the mathematical symbolism for one cycle of alternation in the performance of electrical current. Vertically, it measures the units of voltage per cycle, horizontally, the units of time. Out of this graphic representation has grown the myth that electricity moves in waves, with "short waves" (greater numbers of cycles per second) and "long waves" (lesser numbers).

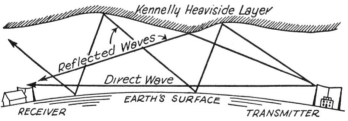

Pursuing the "wave" idiom, this is a crude presentation of the fashion in which radio frequency impulses travel from transmitter to receiving stations. The loss of power in transit is sometimes called "fading," and is believed to occur from absorption of the electrons by the earth and atmosphere.

into the "last radio stage" or carrier frequency, the propagation of this carrier frequency out over a wide area, and the detection and conversion back into sound or image by the receiving set.

At the turn of the twentieth century, inventors were struggling to find instruments which would make all their radio signals coherent, detect them easily, and amplify and modulate them satisfactorily. From the microphone they passed the oscillatory circuit across a spark gap, thereby permitting an induced radiation from the antenna of low-frequency wave trains, subject to interference from many sources. In the receiver, they depended upon the rectifying properties of various crystals to detect the incoming magnetic signal.

It is a curious fact that accidental observation and memory, rather than any direct line of inquiry, led the inventors to a means of increasing the powers of radio to the point we now know. In 1883 there was no electron theory, no understanding of the magnetic flux nor of the nature of wave propagation. But there was a man with an eye for detail and an instinct for discovery unique in human history. Let us imagine Thomas Alva Edison at his laboratory in Menlo Park, New Jersey, examining his wonderful new electric lamp. He observes a carbon mass gathering about the base of the glowing filament leading from the positive terminal of the battery that excites the current. Mr. Edison is not the man to let that pass without a challenge. First, he pastes a piece of tinfoil on the outside of the bulb and taps it in on the circuit. Nothing happens. Then he inserts a plate inside the bulb and between the legs of the filament, but not touching it. He touches the wire leading from the plate to the wire on the negative leg of the circuit. Nothing happens. He touches the plate wire to the positive leg. He is mildly surprised to find the needle on the galvanometer of the plate wire swinging to the right toward the positive leg. Current, somehow, is bridging from the filament to the plate. This is one of the great discoveries of all time, but nobody knows it. Mr. Edison calls in J. A. Fleming, his technical adviser, but Fleming gives up. Nobody has heard of the electron, or its powers of escaping from a current conductor. Mr. Edison is overworked installing electric light systems so he puts his little experiment on the shelf. Only a few scientists continue to tinker with the "Edison effect," in wonder as to why it should occur. What a pity that theory has lagged behind discovery! Had Edison known what a field of operations he had opened up, we might live in a different world today.

On November 21, 1932, Edison effect lamps were inserted in a radio set at a demonstration by the National Broadcasting Company, in New York, and the program continued perfectly well. Edison had created, without knowing it, the one instrument needed to organize the action of the radio wave for coherence.

While the wizard of Menlo Park went on with his other practical experiments, the mathematicians worked for an explanation of what he had done. In 1899 J. J. Thompson announced that the Edison effect was accomplished by the passage of the then recently defined electrons from the heated filament to the plate. Since it had always been the theory that electric current flowed from positive to negative terminals, this action from negative to positive poles was a source of confusion.

At any rate, early radio was at a standstill until Fleming, in 1903, remembered the Edison effect of twenty years before.

By then it was understood that planetary electrons, rotating around the protons of any material substance, do not ordinarily fly free from their orbits without external pressure. However, upon application of heat, some of the unattached electrons acquire sufficient kinetic energy to escape and drift on to be absorbed in atoms having an electron deficiency. The escape of electrons from heated matter is called "thermionic emission," and Fleming set about to make a "thermionic valve" by means of which he could regulate electron escape from filament to plate. But in the place of Edison's simple lighting filament he constructed a more sensitive electrode and termed it a cathode. The plate, with its positive bias for attracting electrons, he called the anode. Electrons, flying loose from the cathode as it became heated from one current, would permit amplification and modulation, according to their quantity, of the current passing through the anode. Thus two electric currents could be brought into tangency, and the linking of oscillating circuits between audio and radio stages could be effected. Fleming called his device the "two element thermionic valve." It was a wonderful instrument, but it didn't last.

Within three years, the inevitable happened. Another man conceived the idea of modulating the flow of electrons from cathode to anode in terms of an independent current. Lee De Forest, in 1906, announced the "three element tube" having a screen or grid between cathode and anode which would impede or accelerate the flow of electrons according to the amount of electromotive force applied to the screen. This could be accomplished because electrons from the cathode could be trapped within the electrostatic field of the grid by giving it a "positive or negative bias." Here was real magic.

De Forest gave his invention the formal name of "Audion" and called it a tube, but properly speaking it is a thermionic valve modulating electric currents. His original instrument contained three elements, cathode, grid, and anode. The thermionic valve today may have as many as five elements within it to govern the behavior of electrons, but basically it is still the Edison effect lamp.

A little giant, the Audion has been called. It is so giantlike that when a President speaks in Washington, the sound of his voice by the mythical fireside is intensified by Audion amplification something like 3,000,000,000,000,000,000,000,000,000 times to carry it across the nation. Such an instrument would seem to be the sort of thing one man would prize highly to sell, another to buy. So dynamic is its effect upon our society that we have not even been able to give it a proper name. A summary statement of the effect it is having upon our institutions and technology has been put most concisely by an agent of the organization which finally made the Audion available commercially, after De Forest had been prosecuted as a faker and threatened with jail for trying to sell stock in his invention.

The invention of the vacuum tube as a satisfactory amplifying, modulating tube . . . opened up the door through which have come not only all of the things which have created this [modern electrical] industry, but likewise all of the things which have created the problems which are confronting this [governmental regulatory] commission and the industry at this present time.

As chief of the great Bell Telephone Laboratories and a vice-president of the American Telephone and Telegraph Company, Frank B. Jewett has a right to speak with authority.

Before we consider how the inventors learned to modulate electricity in terms of varying light as they had learned to do with sound, let us understand just how a scene is caught in television. A great deal depends upon that weakness of the human eye known as "persistence of vision." We retain a mental picture of a scene for about a tenth of a second after we actually see it, so that consecutive scenes, shifted every tenth of a second, tend to appear as a single "moving" picture. It is upon this persistence of vision that Hollywood depends for its illusions. The cameramen shift their fixed scenes at the rate of sixteen per second to eliminate all traces of flicker for the normal eye. With them and with the television engineers, each separate scene is called a "frame" as it appears through the camera's lens.

In the case of Hollywood, the frame is caught instantly and as a whole upon the light-sensitive surface of a celluloid film. In television that is not possible. The frame must be subdivided into smaller units for "scanning." In television, then, one sees not a series of whole frames but of smaller picture units. Imagine you are looking at the frame through a transparent checkerboard. Start at the upper left corner of the board and peek through a single square. Move your eye one square to the right, then another, and another, to the end of a row. You have "scanned a line," in television. Then peek by squares across to the left, back to the right and so on until you have "scanned the frame." The perfection of the completed illusion depends upon the number of lines to a frame, the number of frames scanned per second. To avoid fading and flicker, lines are "interlaced," not scanned in consecutive order. Good 1938 television involves scanning at the rate of thirty frames, of seven hundred lines each, per second.

Television research began long before invention of the thermionic valve. In 1873, three years before a patent was asked for the "speaking telephone," it was discovered that one of the simple chemical elements called selenium was sensitive to light. The structure of carbon is such that the pressure of sound waves upon it will jostle its atoms about and accelerate the passage of an electric current between them. The pressure of light waves was found to have the same effect upon selenium. In the dark it would offer strong resistance to electron movement but would subside promptly upon exposure to light. It made a good "light microphone," and valves containing it within their electrical circuits were called "photoconductive cells." The problem remaining was only that of breaking a frame up into units which would fall successively upon the selenium, and of translating them back into light at the receiving end.

The matter of converting electrical energy into light was disposed of when it was found that electrons, propelled into certain gases, would knock their atomic structures askew, "ionize" them, and set up a glow. The gas responding best visually was found to be neon, which would give off a pinkish sheen from ionization. In 1884, the German, Nipkow, arranged a spiral series of holes around a disk so that one glancing through each of the holes would be bound to see all of the scene framed beyond. Spin the disk, thereby modulating the intensity of light falling upon the selenium in the photoconductive cell, and a frame is scanned. Such was the crude televisor.

At the receiving end there had to be a neon illuminating valve, another scanning disk exactly like the one before the photoconductive cell, and there had to be exact synchronization of the two disks to insure that the flicker of the neon tube would re-create before the eye an illusion equal to that cast upon the selenium at the transmitter. That sort of crude television was possible by means of a wired circuit long before it was learned how to broadcast by electro-magnetic waves of extremely high frequency, the so-called "short-wave" radio transmission now common.

Of course, the early television was extremely poor in picture quality, hard on the eyes, and limited in subjects for broadcast. Other mechanical systems than the Nipkow disk came into use as equipment improved. Elements even more sensitive to light than selenium were discovered, such as caesium, barium, and strontium, the so-called alkali metals. These, it was found, would give off electrons upon being exposed to light, just as tungsten and other materials used in the cathode of the De Forest Audion thermionic valve would expel electrons when heated. Such photoemissive elements were naturally extremely precise and sensitive agents for modulating current flow in terms of light. Hence, the principle of the thermionic valve was adopted, using a photoemissive cathode instead of a heated one, to develop a really powerful "light microphone."

Since this type of valve would respond faster and faster to light rays, mechanical instruments were sought to break each televised frame into smaller and smaller portions, more and more lines. A "flying spot" technique of focusing brilliant light rays upon the scene to be televised led to gearing the movement of the light beam to the turning of a helical series of mirrors, so that as the "spot" moved across the frame each mirror in succession caught a different facet of the whole televised subject. It was found that the light-modulated current could be translated into radio stage frequencies by use of thermionic valve sequences and broadcast short distances by electro-magnetic waves, then detected and retranslated by thermionic valve stages into a final current powerful enough to stimulate a large neon lamp, again flashing upon a helical, synchronized series of mirrors.

But even the finest mechanical system involved a complicated process of synchronizing the gears of receivers by electrical signals with those of scanners. Furthermore, no mechanical means could be found to subdivide frames into a sufficient number of lines to bring out finally a steady, sharp picture clear of flickering and fading. No mechanical method would allow televising of ordinary activities by daylight as would the newsreel camera. Elaborate stage setting, strangely colored lipsticks and face paints were needed to establish actors' features, and intensely brilliant lighting was both expensive and hard on performers' eyes. And so out of the need came the invention, the "cathode ray tube" of modern television, which frees the scanning system of mechanical moving parts and permits a viewing screen that reflects a clear, steady picture.

To understand modern television it is necessary to realize how a cathode ray valve operates. The bulb is pumped as nearly clear of air as possible, so that it has an extremely high percentage of vacuum. The greater the vacuum the less impedance of electron movement that is to ensue. When a current is passed within the bulb from cathode to anode, a glow appears in the end away from the cathode and it seems that some kind of faint light ray is stemming from the cathode. Actually, this is just the electrons cast off from the cathode and bouncing against the glass before streaming back to collect on the anode. They behave so wildly because there is no air friction to slow them down. It has been found that by magnetic coils it is possible to deflect this electron stream and to aim it, as one would aim a stream of water from a hose. The aimed stream of electrons is the instrumentality of high-definition 1938 television. Two techniques of aiming are well known to science and will be the basis of plenty of lawsuits—in fact, have already begun to be so.

One type of electronic deflection is known as the "Farnsworth image dissector," an invention of a young American, Philo Farnsworth, of whom more later. In this device, the frame of the picture to be scanned is focused through an ordinary camera lens upon a translucent, photosensitive cathode. As the whole scene falls upon this cathode at once, electrons go gyrating backward through the tube in myriads toward the anode. The focusing coils around the outside of the valve straighten them out and move them in orderly, parallel lines, so they end up raining upon the

INVENTING A NECESSITY 33

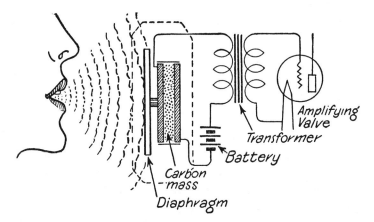

When one speaks before the microphone the sound waves strike against a diaphragm, agitate the carbon mass behind it, and cause current to flow through the microphone in amounts varying according to the frequency of the sound wave.

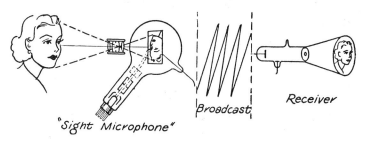

As in sound, so in sight broadcasting. The difference here is that modulation occurs upon the light-sensitive plate and is converted back into the shadow of substance by propulsion of electrons upon fluorescing material at the broad end of the receiving valve. May we be forgiven the phrase, "sight microphone"?

anode with a distribution of electrons corresponding to the distribution of light intensity upon the cathode. There is a small aperture in the center of the cathode which corresponds to the hole in the scanning disk.

The picture is scanned by running electric currents through the focusing coils so they displace the electron image at the anode in a systematic fashion to allow a constant stream of electrons to pass through the aperture and fall upon another "output" electrode which amplifies their effect. This procedure is somewhat like a patterned rain of bullets which would cut a design on a wall, except that the bullets go through a single hole and recreate the pattern elsewhere. In the case of the electrons it is not the hole that moves, but the bullets, for they are deflected systematically by the magnets. From the output electrode, the faint stream of electrons is converted into a powerful radio wave.

Farnsworth's chief rival in the development of cathode ray scanning is Vladimir Zworykin, a Russian, now in the United States. In Farnsworth's instrument the electrons fly off the photosensitive translucent cathode, are deflected systematically by magnetic coils en route to the anode and sent successively through an aperture to the output electrode. Zworykin's device begins with a common ray of electronic beams fired from a cathode crater at the lower end of the bulb but the beams never get to fluoresce against the glass. Within the bulb there is a screen made up of a mosaic of tiny segments of caesium or other photoemissive element, with each mosaic insulated from the ones surrounding it.

Upon this screen a lens focuses the scene to be televised: the frame. Deflecting coils cause the electron beam from the cathode to scan this screen mosaic according to a definite pattern which fixes the number of lines per frame, the number of frames per second, of the finished picture. As the light from the televised scene falls upon the mosaic, electrons are lost, of course, by photoemission from the globules of caesium and the metal back side of the screen, made of mica or other insulating material, builds up a positive charge equal to the light intensity. But when the cathode beam moves across the screen it makes up the electron deficiency in each globule caused by photoemission, thereby discharging the globule of its imbalance, and sending out an electrical impulse from the metal back of the screen which had served as a condenser of the originally photoemitted energy. The output electrode, as in the Farnsworth system, modulates the current passing through amplifying valves so that at the last radio stage an impulse of high frequency intensity is sent out from the antenna.

By whatever name you hear it called,—oscilloscope, kinetiscope, kinescope, or otherwise,—the valve used in the modern television receiver is just a cathode ray tube, with a fluorescing material coated upon the end of the bulb to give the highest possible luminosity to the glow caused by the collision of the electrons being turned in their flight from the cathode. The electron beam scans the end screen in an orderly manner because the magnetic coils deflecting it are synchronized by means of a radio wave with the movement of the scanning beam in the iconoscope or image dissector. The degree of brilliance is maintained by the voltage applied at the anode of the receiver tube. Of course, the final picture definition and clarity depend upon the number of lines per frame, the number of frames per second scanned by the televisor. The immediate problem of television is how to enlarge this cathode beam picture upon a screen sufficiently large to insure the greatest possible eye ease, and to reproduce in natural colors.

4. Wires Versus Wireless

TWO MEN MAY THROW ANOTHER DOWN AND PIN HIS ARMS until he agrees to obey the rules. A government may suppress revolt, enforce concepts of property, value, taxation, crime. But it has not yet learned to suppress the instinct of idle curiosity, to murder thought, to predict finally that only so much and no more may be expected of an invention. On that account grave judges, wise holders of great money values, learned counsel, and stockbrokers with margin accounts on their hands are most painfully embarrassed from time to time.

Law is based on understanding, property upon things real and tenable, income upon promises that must be kept or something valuable forfeited. What does an electron know of these things? One set of men learns how to make it perform, always with the speed of light, always fleeting; another set tries desperately to make the performance worth something in money. For a little while rewards are great.

But new uses come. The show moves elsewhere. And the investor, having paid for a performance he is not to enjoy, fumes for a time in the empty theater before going out for his lawyer, his glass of cyanide, or fresh money for a new try to keep up with the fun.

It is quite obvious that DuFay, Edison, Fleming, De Forest, the busy thousands whose intellects have gone into the building of modern communications systems, wanted something out of life other than just the pleasure of making electrons do tricks. But what? Not money, merely. Of the forty men who did pioneer service of a major nature to bring radio activities up to a reasonable standard of technical performance, only two ever received any appreciable monetary reward. One died with an estate of less than $150,000 and the other went bankrupt.*

The audio frequency was the basis of one great fortune and industry in the world, that centering around the telephone, but the actual inventors of the instrument were never greatly rewarded with money in comparison with the staggering sums the inventions have earned. Great laboratories, in recent years, have been better disciplined, more safely operated, than were the little shops of times past. Knowledge is better organized, but profit still goes to others than the inventors. We cannot pause here to shed tears for them and we must pursue the history of their achievements.

Out of the audio current emerged the radio, or high frequency carrier current, and so there was developed our great wireless industry—and with it a technical jargon that will probably be centuries in disappearing.

For instance, in the early days it was commonly said that a set operated on a wavelength of so many meters, meaning that the time interval between one pulsation of electromagnetic waves and the next was such that the first would have gone, say, five hundred and fifty meters into space before the next could follow off the transmitting antenna.

As the instruments of broadcast were improved, this time

* Guglielmo Marconi and Lee De Forest.

interval came to be made briefer and briefer, and so it became very difficult to measure wave intervals in terms of meters. The fashion changed, then, to speaking in terms of frequency of completed cycles of pulsation, from zero to zero voltage. These cycles speeded from a mere fourteen a second past fourteen thousand and on until the engineers were not afraid to say they might some day be able to send out a million impulses a second. This led them to introduce another omnibus word into our common language. They found it necessary to speak in thousands of cycles, and so they soon fell into the habit of saying a set operated on so many kilos of cycles, kilocycles—"has a frequency of 560 kilocycles." The cycles came faster and faster, until today they speak of and deal in megacycles, millions of cycles, micro-waves, extremely minute separations of waves in motion. The succeeding impulse off the antenna is closer and closer to the one before, an inch, less than an inch, a tenth of an inch. Some day we may find electron and electron flying off the transmitter into space as closely together as they do in an atom. What, then, will be the nature of a broadcast? David Sarnoff, president of the Radio Corporation of America, has collected the views of his engineers on the future of radio and stated them thus:

This expansion of the useful radio spectrum has only begun. Beyond the ultra high frequencies lie the microcycles, frequencies that oscillate at the rate of a billion cycles a second, wave-lengths measured in centimeters instead of meters.

Future developments in micro waves may well prove revolutionary. In the past, radio operations have been confined to a limited part of the radio spectrum. Once we have conquered these micro waves, we shall have opened a radio spectrum of almost infinite extent. Instead of numbering the desirable channels in a few scant thousands the radio art will put millions of frequencies at the command of communications services of every kind.

When that day comes—and I have no doubt that it will,—there will be frequencies enough to make possible the establishment not only of an unlimited array of mass communications services, but of an unlimited number of individual communications connections.

In that day, each one of your millions of citizens may have his own assigned frequency to use wherever he may be. Step by step we are working toward that far off goal.[1]

What sort of talk is this?

Sarnoff is no madman, but the responsible commander of one of the world's greatest corporations. Back of his words rest engineering knowledge, fortified by finance, law, and confidence. What is this radio spectrum? We shall define it in detail, presently. What kind of men and women will we be, when each of us has his radio frequency on that spectrum? Will we have long ears and big eyes, little teeth and withered hands? Maybe so. But there is at least a faint sign that the man of our time may know such things without having, necessarily, to submit to physical decrepitude. One American watch company has applied for a frequency to operate a watch upon a radio signal, the watch to be worn and carried wherever the owner may wish.[2] That, at least, is a beginning. We have radios small enough for cars. And television programs are broadcast between ship and shore, plane and earth.

Sarnoff's words are promise of a wonderful world. But here is a dissonant note:

Twenty years ago, or even fifteen years ago, the bridge of the radio spectrum was narrow as we envisage it today.... But while it has increased very considerably, it looks to me pretty much as though the need for the services in numbers and uses or services in number or in kind, have pretty well kept pace with the progress of our proven knowledge, so that whatever we may think of in the long run or distant future, such as Mr. Sarnoff envisioned for us, certainly today as I see it radio is still a service, is still a means of transmission in which the number of channels is distinctly limited.

They may be large; they are very much obviously larger than they were a few years ago, but it is not a mode of transmission as yet in which we deal with it on the basis that we have an infinitely large number of channels which can be used ad lib.

As I see it when the shoe begins to pinch, as it obviously is pinching in some sectors at the present time, the problem which will confront the [governmental] Commission will be in the last analysis a problem of relative merit of modes of transmission.

On one hand, where things like the radio are unequivocally indicated as the sole or practically the only way of giving service, that prevails; but if it comes to the proposition of the thing in which it is obviously an alternative, the question becomes an economic one, and the question as to whether we should use up a portion of our limited resources for a thing for which there is an alternative, will have to be determined by the question as to whether the differences in costs are such as to justify such an expense.[3]

Who is this slightly incoherent gloom thrower? Like Sarnoff, he, too, is a man of authority, high placed in a world of science, money, law, and power. His name is F. B. Jewett, and he is the selfsame one who spoke so feelingly on the

De Forest Audion valve. As the director of the Bell Telephone Laboratories and a vice-president of the American Telephone and Telegraph Company, his words must be considered to have important inferences. One inference is clear. He disagrees with the concept of an unlimited spectrum.

Are Jewett and Sarnoff about to lock horns like buck deer in the springtime? Sarnoff and Jewett are the mere spokesmen of opposing myriads of dollars, brains, laws, and electrons. They represent opposing interests of mastodonic size, one wanting to use electrons by way of the spectrum, through the free and as yet untaxed air, while the other, representing a great fortune invested in wires, stands ready to fight for confining electrons to them. If they meet in battle disaster is as likely to overtake the winner as the loser.

We cannot escape the feeling of reckless disregard of consequences as shown in Mr. Sarnoff's view of Mr. Jewett's carping:

> From time to time, there are suggestions that it is the duty of the Federal Communications Commission to protect the wire services against the encroachment of radio.
> Even if the Communications Act which created your commission had not prohibited such an attempt by saying that your commission shall "generally encourage the larger and more effective use of radio in the public interest" such an effort would be a futile one.
> Any effort to stop the progress of a new art in order to protect the existing art is bound to be futile.
> Such a step would be contrary to the spirit of the country, contrary to the modern spirit of progress, and contrary to the whole experience of radio. For radio itself deliberately obsoletes [sic] today what it built yesterday.[4]

An admirable spirit, you would agree, and one that admits institutions must surrender to technical advance; that men must adjust themselves to the electron, not expect the electron to submit its powers, once discovered, to be hidden away and never used.

5. New Public Property

IN SURRENDERING INSTITUTIONS TO TECHNOLOGY THERE ARE difficulties which may cow even Mr. Sarnoff, but if he triumphs over them he may become a Caesar such as never was before in this world.

Electro-magnetic waves know nothing of commissions, fair-return-on-investment, in-the-public-interest, Constitution-of-the-United-States, or even of the so-called "radio spectrum." They do not even know one program of entertainment from another. If they are sent out from a transmitting station they travel so far, according to the voltage pressing them. They produce specified effects, according to the frequency of their emission, and they are received by all instruments attuned to catch them.

Here then is implied a battle more immediate than that hinted by Mr. Sarnoff and Mr. Jewett as coming between radio and wire transmission. Radio programs in our time are valuable in terms of money, and in terms of power over other people's minds. Piracy clearly is to be expected if we cannot govern the sending and receiving.

To describe the behavior of the electronic impulses in the old way, the ether may be infinite, there may be room for millions of individual frequencies as Mr. Sarnoff says, but on a given day the ether is not infinite but limited and

NEW PUBLIC PROPERTY 45

divisible according to the number of frequencies then actually usable.

Two stations broadcasting within range of each other are going to create confusion for the listener, obviously. How, then, to accommodate all who seek to broadcast? This problem, universally, has been attacked by governments. The "ether," they have proclaimed, is a public property. No man can own it, no man can drive a stake in it, mark off boundaries, and declare: "This is mine. I have found it and I am going to keep it, by God and my right, as long as I pay my taxes."

Instead of setting up private property concepts which are attacked the moment two stations on the same frequency broadcast within range of each other, most governments have tried other means. Within their domestic boundaries these governments have claimed radio activity lock, stock, and barrel for their own. In many cases they retain outright ownership of the broadcasting stations. Individuals may make apparatus, but none may send out programs except under direct supervision of the bureaucrats in power. No problem of internal regulation exists in such stations.

Their difficulty is in maintaining prohibitions against broadcasts from foreign stations that bring "false propaganda" to the ears of Government X's notoriously happy subjects. In an attempt to eliminate this ethereal anarchy and subdivide new domains opened by improved scientific technique, international conventions are held at stated periods. Early in 1938 at Cairo, Egypt, technicians in law, diplomacy, and engineering set out to adjust agreements concerning who should operate on this frequency, who on that, and according to such and such basis of voltage and power to carry over long distances, in order to prepare the way for general television operations and to curb international piracy in sound radio.

Their problem has long since become vastly complicated. International propaganda by radio is a technique of the age. It is growing in use everywhere. For years it has been commonplace in Europe. The United States has just entered into it. We cannot repeat too often that the revolutions, wars, and other infections of the popular mind which have been noted in recent years have been attributed by ardent radio propaganda enthusiasts to failure by the United States to defend democracy in the Western Hemisphere with the same means by which followers of the other faiths attack it.

And so, as the technique of the engineers permits broadcasting to reach out further and further, the difficulty of preventing international piracy and chaos among the electrons becomes much like the situation current in the one country of the world that has allowed its citizens to play and profit with radio activity on a large scale.

In the United States private citizens, corporations, even the agencies of the Government itself, are required to conform to regulations as to who shall use this frequency, who shall use that, whose station will have thus and such territory and whose shall have another.

The aggregate sum of all of these allocations is vaguely called "the radio spectrum," a misleading phrase which suggests something connected with scientific analysis of light or the electro-magnetic spectrum of radiations not visible to the eye and varying in frequency from the long slow ones

NEW PUBLIC PROPERTY 47

of eighty-five per second to the gamma rays, pulsating at speeds of 100,000,000,000,000,000,000 per second.

The "radio spectrum," as used in the jargon of broadcasting, is simply the listing of frequency allocations to operating stations. It is the Ark and Covenant of the twentieth century radio electronics in the United States. Its high priests and keepers are known as the Federal Communications Commission, whose members are seven, the allegedly lucky number.

Should not they be among the wisest men of our time, in view of what his Cabinet officers have told our President about the dangers in television? The most thoughtful, the most eager to expand that radio spectrum as Mr. Sarnoff urges, by encouraging the art of electronic radiation at all costs? Do they encourage it? Can they?

They are not free.

Once a frequency is allocated, it is used at great expense, and customarily with great profit. Nobody yet has relinquished his position in the spectrum voluntarily, and solely because another desired it for a better, different purpose. Yet if that better purpose is to be achieved, the commission must take away, even as it gives. It must deprive as it grants, for the spectrum is finite. Its divisibility is always known, and so far more apply for frequencies than there are frequencies to be had. And the depriving and granting lead to painful, often unsuspected, conflicts. To understand the nature of these we must return to technology.

The radio spectrum is not like a mosaic, a series of clearly defined squares and straight lines, upon which can be shifted licenses like counters on a checker board. Because of the way electrons behave, it is like a plate of spaghetti, lacing and interlacing, winding and weaving, built of compromises and adjustments. To increase power on a given frequency or to decrease it may have infinite effects upon electronics and upon property.

6. The Inadequacy of Law

THE RADIO SPECTRUM IS FINITE, LIMITED. TRY AS THEY WILL, the engineers have not yet reached the standard promised by Mr. Sarnoff; that is, they have not provided a technical basis for unlimited operations free from interference.

And ever since radio has been in existence, government has sought earnestly for ways and means of dividing the fields of operations most equitably. The problem was never put more succinctly than by Louis G. Caldwell, former chairman of the American Bar Association's committee on radio law and former member of the governmental regulatory commission:

Another message that the facts and principles of radio brings to us is that without rigid government regulation, you are not going to have any radio communication at all.

If the individual has the determination of whether he will or will not use a radio apparatus there is going to be chaos and anarchy in the air, so the government must not only have the right to determine who shall be in the field, but must have the most extensive rights to regulate those that are in the field, both as to their technical operations and to see that they are fulfilling their duties under the test provided by the law.

The broad problem of the [governmental] commission is to apply this test which you have set for it, that of public interest, convenience or necessity. We know, of course, that phrase has a public utility history. It reasonably presents some new problems because in no existing public utility has there really arisen the necessity of putting anyone out of business that is already in it. There has been natural room for all existing concerns in other lines of business.

In radio it is going to be frequently necessary, I think, to put someone out of business from time to time to make room for someone else.[1]

A grim task, this, and one not entirely understood by all who may be required to decide between life or death for the radio operator. Consequently, we find recurring conflicts of opinion by authorities. The United States Court of Appeals for the District of Columbia, in examining this premise granted by Mr. Caldwell that "it is going to be frequently necessary" to put someone out of business, only recently warned that it was by no means in agreement with the contention frequently urged that evidence showing economic injury to an existing station through the establishment of an additional station is too vague and uncertain a subject to furnish proper grounds of contest. The court held that in any case where it is shown that the effect of granting a new license will be to defeat the ability of the holder of the old license to carry on in the public interest, the application should be denied "unless there are overweening reasons of a public nature for granting it."[2] Just how could a judge or anybody else have the heart to say that television is an "overweening reason of public nature" for putting the present familiar sound radio out of existence, in view of all the good services it now performs?

Yet somebody must decide, since engineers indicate that

such a death may be necessary. The United States has delegated this delicate problem of effecting the survival of the fittest to the seven members of the Federal Communications Commission, especially authorized to determine public interest, necessity, and convenience in radio and other forms of electronic communication. The commissioners are aided in their work by large corps of engineers, lawyers, and clerks. They make an imposing array of public servants, and they have come to power after a long and not altogether glorious struggle between the legislators and the electron. It is a struggle that began with the advent of the twentieth century, when shipping lines began to use the wireless telegraph. Several companies were then making communication equipment and furnishing service between vessels at sea and points ashore. But in their all too human way, the operators of one company would refuse to deal with those of another. Even distress signals were ignored or uttered falsely to plague rivals. Ships were boycotted and refused reports on the weather, the prospect of cargo, or other vital knowledge solely because they were not using the equipment of the organization from which they were seeking information.

In 1906 Kaiser Wilhelm of Germany, enraged by English and American abuses of German equipment and stations, forced a treaty according to which all wireless stations were bound to connect with each other upon demand without consideration for differences in systems or instruments.[3] This was the first fiat of any kind to bring together the exploiters of the electron upon a communal basis. It set a precedent that has grown in power with every new attempt at legislation for proper use of the radio.

The United States, by 1910, was sufficiently concerned

about the use of the wireless for Congress to require that certain classes of ships be equipped with it. Licenses of operation were granted by the Secretary of Commerce and Labor, but not according to any elaborate formula.[4] Two years later the impact of events demonstrated how a single incident can turn the tide of a life. In 1912, when the *Titanic* sank, an operator for the American Marconi Company at a station on top of Wanamaker's store in New York City was the only man in the United States to catch the message. At least, he was the only one who knew what to do about it.

That radio operator sold the news of the *Titanic* disaster to the Associated Press and turned the money over to his employers. He recognized then the value of news and he has demonstrated since that his sale of it was nothing haphazard, but an instinctive action. The name of the operator was David Sarnoff.

The sinking of the *Titanic* had other important effects on radio. It caused Congress to ratify the 1906 treaty of Berlin in a hurry and send delegates to a conference in London.[5] Then Congress passed a new law making it mandatory that the Secretary of Commerce grant licenses to stations transmitting information in interstate commerce, but neglecting to specify that his action must be based on the public necessity or convenience. It gave him no authority over content of messages, duration of license, technical standards, or title of ownership.[6] The control of the electron by the Government of the United States was only implicit; and the inadequacies of a law based on such tenuous stuff made for many heartaches and frenzies of investors later.

The Attorney General of the United States advised his

cabinet colleague in a formal opinion, soon after enactment of the 1912 law, that he was without discretion to withhold a license to any citizen of the United States who should apply under that Act; that anybody could go in business if he had the will and the money.[7] But the busy inventors soon showed all that up as so many silly words based on ignorance and misunderstanding of the real problem.

De Forest's audion tube, Alexanderson's alternator, the Poulsen arc, the Fleming valve, the Armstrong regenerator —these and a myriad other devices to wring contortions out of the electron began to appear. The then known spectrum was choked by interfering broadcasts and the courts began to hear new language which could neither be understood nor called contemptuous by the learned jurists. Interests conflicted, and the electrons refused to serve in a world devoid of co-operation. These circumstances led to an explosion in 1923, when an applicant who had been refused a license brought a suit before the United States District Court of the District of Columbia intending to force the Secretary of Commerce to conform. The matter came eventually before the United States Circurt Court of Appeals for the District, and the decision of that court was accepted hopefully by government and industry as establishing the rule of law over the vagaries of the electron.

It held that while the Secretary was bound to issue a license as he had been advised by the Attorney General, he could place restrictions upon its use as to power, hours of operation, and other technical qualifications.[8] This judgment had no basis whatever in the Constitution, as subsequent events clearly showed, and represented an unquestionable effort to legislate by judicial fiat.

On the basis of this decision the Secretary of Commerce began to issue licenses limited as to the important engineering aspects with which we are familiar, and worked out what amounted to a sort of rudimentary spectrum, but still he could not catch up with the fleeting electron.

The Secretary defined eighty-nine basic frequencies, which by 1925 were jammed with five hundred and seventy-eight broadcasting stations, each fighting for increased hours of operation, increased power, and superior frequency. The air was turbulent with interference and piracy. A violation of the Commerce Department's regulation made one liable to a fine of twenty-five dollars—no punishment at all for the operators who were finding in radio a playground for wild money comparable to just one other of our time, the movies.[9] A further test of the 1923 decision was inevitable. It came at Chicago, early in 1926, when the Zenith Radio Corporation set its station, WJAZ, to full time operation on an unauthorized frequency.

The U. S. District Attorney promptly sued out a writ of injunction based on the law of 1912 and the decision of 1923. As promptly, the U. S. District Judge set it aside and declared invalid the decision of 1923 upon which the spectrum had been built on the grounds that no such powers were stated or implied in the 1912 act, and that the act itself was of questionable constitutionality. The government did not contest the new decision and let the 1923 principle of regulation lapse.[10] Chaos ensued.

Until the Secretary of Commerce could bring operators together in a "gentlemen's agreement," piracy and interference were standard hazards for all who ventured into the business, but two hundred new stations sprang up just the same. In spite of all the difficulties, radio became immensely popular with the general public. "Coon Sanders' Nighthawks" were the most popular dance orchestra of the hour, and a little man named Snodgrass played his way out of jail in Missouri to the tune of "Three O'Clock in the Morning." Boys built receiving sets according to mail order instructions with all kinds of equipment.

In October, 1926, the issue of property right came up for a belated and insufficient test. Two Chicago operators clinched concerning a frequency, and one asked a chancellor in the Cook County Circuit Court of Illinois to restrain the other from interference. The learned jurist, drawing upon the precepts of the English common law, said that the one who had been using the frequency longer had the prior claim. He indicated that in his view one might even stake out a permanent hold in radio and apply "no trespassing" to the heavens, as it were.[11]

In December, 1926, there was in existence a special committee of the American Bar Association which published a report holding that existing stations had a property right in the use of the ether and recommending that Congress provide compensation for any station which, under a new law, might have to cease operation.[12] That opinion is a fair illustration of the trouble a committee of lawyers who base far reaching conclusions on snap judgment and inadequate knowledge of the problem before them may cause if they are able to alter the judgment of persons in positions of public responsibility.* Had the lawyers pondered things a

* In 1936, a committee of counselors for the American Liberty League declared the Wagner Labor Relations Act unconstitutional, and many employers promptly took their advice as a basis for ignoring it; but the Supreme Court did not. It held the Act lawful.

bit more deeply they would have realized that the government really was about to hand them a grand bonanza field for a new kind of "law" practice.

In 1927 Congress passed a new radio act which provided that a Federal Radio Commission of five members, appointed by the President with the advice and consent of the Senate, should determine frequencies and power, grant licenses for limited periods of time or deny them, and do all the other obviously necessary things for development of an adequate spectrum.[13] Who but lawyers could appear before the commission and help it to interpret the new law? Though its problem was a problem of engineering and allocation of physical affairs, the solving was a solving by legalists.

By injunction, ruling, guess, and prayer the commission carved out a policy of determining public convenience and necessity and finally, after weeding out many stations and establishing for itself a reputation for being less than perfect, began to rebuild the spectrum. But even as the commission hacked through the legalistic undergrowth, the engineers were racing ahead. The Bell telephone system opened its wires to radio on a nationwide basis in 1926, just after the breakdown of the law. Then came the linking of stations in chain broadcasting and all the science and artistry of selling merchandise from coast to coast as an excuse for amusement, drama, and music. The amateur radio operators, who had been pioneers in the development of the art from its inception, were driven from the air for a while, as nearly every frequency then extant was given over to commercial broadcasters. But the amateurs began explora-

THE INADEQUACY OF LAW 57

tion in high frequency ranges. Their discoveries opened up new segments on the spectrum and began to excite belief that it was wholly practicable for one to see programs as well as hear them.* Television, which had begun as laboratory dream stuff so many years before, was, by 1928, established as practical. The amateurs caused the opening up of a whole new universe for radio. And in 1934 Congress found it necessary to broaden the law again.[14] The art of electronic communication was rapidly approaching a synthesis in methods. Something had to be done to bring order before the arrival of television in general public use.

No statute in the United States Code reads more grandly than the Federal Communications Act of 1934. Here would seem to be modern legislation in the enlightened vein. Section I of the general provisions states that the Communications Commission of seven members is created

... for the purpose of regulating interstate and foreign commerce in communication by wire and radio so as to make available, so far as possible, to all the people of the United States a rapid, efficient, Nation-wide, and world-wide wire and radio communication service with adequate facilities at reasonable charges, for the purpose of the national defense; for the purpose of promoting safety of life and property through the use of wire and radio communication; and for the purpose of securing a more effective execution of this policy by centralizing authority heretofore granted by law to several agencies. ...

* We regret that we have not been able to devote a whole book to discussion of the amateurs in American radio. There are today approximately forty-seven thousand amateurs licensed to operate, and they constitute one of the most fruitful of all research groups.

Power is given to regulate in minute detail every wire or radio common carrier of messages between states and between the United States and foreign nations.

It is the purpose of this Act among other things to maintain the control of the United States over all the channels of interstate and foreign radio transmission; and to provide for the use of such channels, but not the ownership thereof, by persons for limited periods of time under license granted by Federal authority.

The commission is told to classify radio stations and to prescribe the nature of service to be rendered by each class and each station. It is empowered to act *in the public interest, convenience, or necessity* by granting, suspending, altering, or revoking licenses. It is expected to "study new uses for radio, provide for experimental uses of frequencies, and generally encourage the larger and more effective use of radio in the public interest."

By the commission's own order, every operator's license must be brought in for renewal every six months, but the law makes sure that the public grip on radio will be preserved in any case by providing that no class of license can be granted irrevocably for a period exceeding five years, and that no license for a broadcasting station shall run for more than three years without tests for renewal.

No alien or his representative, no foreign government or corporation organized under a foreign government, nor any corporation of which any officer or director is an alien or in which alien interests hold more than one-fifth of the capital stock, shall be granted a license. This precaution is even extended to prohibit the licensing of any agent of such alien interest, or holding corporation with alien officers, or alien ownership of so much as one-fourth of its stock.

The commission is specifically directed to refuse a license to any applicant finally adjudged guilty in a Federal court of so much as attempting to monopolize unlawfully radio communications or the manufacture and sale of radio equipment, or attempting to use unfair trade practices. Of this, more later. There are serious strictures upon the commission to preserve competition in commerce.

Two passages must be read in detail for appreciation of the earnest effort by Congress to conform to the classical principles of democracy in radio and still keep a governmental grip upon this novel device for reaching millions of people at once.

If any licensee shall permit any person who is a legally qualified candidate for any public office to use a broadcasting station he shall afford equal opportunities to all other such candidates for that office in the use of such broadcasting station, and the Commission shall make rules and regulations to carry this provision into effect: provided, that such licensee shall have no power of censorship over the material broadcast under the provisions of this section. No obligation is hereby imposed upon any licensee to allow the use of its station by any such candidate.

And:

Nothing in this Act shall be understood or construed to give the Commission the power of censorship over the radio communications or signals transmitted by any radio station and no regulations or conditions shall be promulgated or fixed by the Commission which shall interfere with the right of free speech by means of radio communication.

No person within the jurisdiction of the United States shall utter any obscene, indecent, or profane language by means of radio communication.

Private messages, such as telegrams and telephone conversations, are declared inviolate; and in addition to forbidding interception or listening in on any of these, the Act provides for as much as a two-year prison term or a ten-thousand-dollar fine for violators. The Supreme Court has held that evidence gathered by such wire-tapping cannot be used.

"Broadcasting," says the Act, "means the dissemination of radio communications intended to be received by the public directly or by the intermediary of relay stations."

All hearings, testimony, and findings of the commission are matters of public record, and relief to those who feel they are aggrieved by commission verdicts is provided in appeals in the U. S. courts. Finally, the President is empowered to seize the whole communication system in time of war and use it as the emergency dictates. Nowhere is it stated that the public interest, convenience, or necessity is served by the sale of radio broadcasting time for commercial purposes. That is a presumption by the commission, apparently, which has been the basis for granting licenses to one applicant and denying them to another.

The Communications Act provides plenty of powers, it is clear. But it does not guarantee progress. It is a curious but undeniable fact that the radio industry has thrived and progressed not under routine commission government, but during and just after extensive congressional investigations into the state of competition between the chief participants in the business. These inquiries appear to stimulate latent or withering competition and to spur dominant corporations into demonstrating their proficiency by bringing out new products and new techniques.

7. The Philosophy of the Spectrum

IT IS IMPOSSIBLE TO UNDERSTAND THE PECULIAR PAINS THAT rack the radio industrialist unless you concede him a little obliquity of speech.

He says, for instance, that the Federal Communications Commission has declared the range of the useful spectrum to run from 10 kilocycles to 300 kilocycles. What he really means to tell you is that the commission has announced jurisdiction over all instruments used in interstate commerce to transmit messages by means of ten thousand to 300 million cycles of electro-magnetic impulses per second.

Again, he says that he has just acquired "an F.C.C. license to broadcast at a frequency of 1.7 megacycles on a band 5 megacycles wide." What he means to tell you is that the Federal Communications Commission has granted him a license to operate his machine so that the current alternates at a frequency of 1.7 megacycles per second, and that no other station is licensed to operate on a frequency with a range of 2.5 megacycles greater or less than his own precious allocation.

The reason why other operators are set apart is that if two stations of approximately the same frequency operate within range of each other, reception of one signal is likely to be

THE PHILOSOPHY OF THE SPECTRUM 63

confused by the other. Different types of transmitters require different width bands, to use the radio man's term. But there is one fact of life that none of them can escape: to avoid confusion, everybody in radio must know what everybody else is doing in the way of sending messages. This is because of a natural phenomenon known as the Kennelly-Heaviside layer.

Imagine broadcasting a radio signal and hearing it echoed into your receiver after a second or so. Two scientists by the names of Kennelly and Heaviside performed that little stunt and came to the conclusion that the earth is encased in some sort of atmospheric envelope beyond which radio signals do not pass. They said there must be a roof over the world off which radio broadcasts bounce like rubber balls. Other scientists call this envelope the ionosphere, and say that it not only bounces radio programs back to earth from its inner side but, from the universe beyond, absorbs a strange radio-active hail of "cosmic rays," some of which still drive through to condition the physical world.

The turning of the earth upon its axis, the bombardment of cosmic rays, the bouncing of radio waves off the Kennelly-Heaviside layer, and the mutations of sunspots all affect the radio industrialist. These phenomena condition public interest, necessity, and convenience, the value of common stocks, and the width of broadcasting bands more directly than any man-made regulation.

Combinations of them produce effects that have led us to liken the radio spectrum to a plate of spaghetti, rather than a mosaic of colored bricks. For instance, when the police headquarters at Newark, New Jersey, broadcasts a message to a scout car, the call goes echoing and rattling off through the heavens to be picked up in most unexpected places. Such local calls, as a matter of fact, have been heard plainly in San Francisco, on the Argentine pampas, and in Berlin, though they were so placed on the spectrum that they skipped over near-by areas without being noticed.

The spectrum writhes, then, like the spaghetti, yet it shatters at the touch. If a single frequency is shifted, a single band widened or made narrower, the natural phenomena of radio may bring about disastrous results to commerce. That is why, when the commission announced its intention in 1936 of reallocating positions on the spectrum, Ralph M. Heintz, former president of the Radio Manufacturers' Association, cried out:

> I hate to see anything happen to that portion of the spectrum where large and expensive and high-powered equipment is placed, where large and expensive antenna systems are a part thereof, where one little twitch in any portion of that spectrum makes the whole spectrum shiver from one end to the other.
> Of course we all shiver in our boots along with it.
> So it would seem highly desirable to have things stay just as they are. . . . Let there be congestion rather than do anything about it that might upset other branches of the service.[1]

Mr. Heintz speaks a language which the jurists understand and he speaks for all who have found a foothold in the tight and teeming coral reef of the spectrum. He wants to preserve the status quo and let the advance of technology conform as best it can. But the Federal Communications Commission is charged with promising no radio licensee anything beyond that set forth specifically in the language of his license. And the license says he must prove every six months that he is earning his right to life by good works in the public interest, as well as by his faith in the right of a man to keep that which he has earned by his labors. Who fights for a status quo in radio? Who seeks change? It is a curious fact that we will find one is often the other; that the same man grows a new leg and tears off an arm, so to speak.

The interests in radio are by no means limited to commercial broadcasters peddling the hands of the clock and the songs of girls to makers of dogfood and beauty creams. Nor are they static in number. When the first important international list of radio frequencies was compiled at Berne, Switzerland, in 1928, a total of one thousand seven hundred transmitting stations was reported. By March, 1936, there were twenty-five thousand, exclusive of amateur, ship, aircraft, and portable transmitters, which, while not estimated officially, probably number more than five times that figure today.[2]

In the United States, the domestic spectrum gives an index to a tremendous but for the most part unsuspected business of radio. Approximately fifty-five thousand stations[3] have been authorized. Here is how the spectrum is divided:

Between 10 and 100 kilocycles, there is room for two hundred and seven radio-telegraph channels. But if the same space is converted to low-quality radio-telephony, it permits only fifteen channels. Thus we see the commission is immediately confronted with a problem of selection. This segment also allows six high-speed facsimile channels and only four high-quality telephone channels.

It is best adapted to high-power communication between

fixed stations long distances apart and is therefore considered international in its service range but not wide of scope, for it does not offer room for ordinary sound broadcasting or other special services. Only forty-seven stations in the United States and three hundred and ninety-five abroad were operating in that segment during 1937.

In the medium frequency band, 100 to 550 kilocycles, the overcrowding begins. The 1936 estimate of users was six thousand eight hundred United States stations and about two thousand seven hundred and fifty foreign fixed and land stations, including general governmental; special types, such as foresters and power companies, for inter-office communication; operators of radio beacons and direction finding instruments; aeronautical and airport systems linking planes with the earth; and ship and coastal services. Here, too, are radio typewriters and radio operated bookkeeping systems, by means of which a central office in one city checks accounts in branches around the nation.

What is known as the "commercial broadcast band" comes next, between 500 and 1600 kilocycles. Here travels the electron to bring you dance bands, comedians, political conventions, and the polite urgings of "sponsors" whose "generosity makes this program possible." This is the current but threatened Klondike of the air, the portion of the frontier most rich in immediate cash reward but not yet proved a permanent harvest land. The broadcast band is subdivided into clear channels enjoyed by a few all-powerful stations which are entirely clear of interferences (or competition), regional high power channels, regional channels, local channels, and Canadian-shared channels. In 1937 there were seven hundred and four licensed broadcasting stations using ninety channels in this segment. But there were also six government channels in operation, much to the annoyance of some disappointed private applicants.[4]

The medium high frequency band lies between 1600 and 6000 kilocycles, so we step up in our units of measurement and say it is between 1.6 and 6 megacycles. It accommodates nine hundred and fifty standard channels, including the recurrence of agencies we have met above, such as marine, aviation, police, amateur, point-to-point, forestry, and some other special services. Power companies have also found a place here for field communication between surveying parties. Here television has already struggled for existence and won a partial victory. On May 13, 1936, the commission decided to move television out of this portion of the spectrum entirely and give more space for police, aviation, and some other special services.

Purdue University, holder of an experimental license, made strenuous objections, on the basis that only between 2 and 2.85 megacycles could television be broadcast into rural areas. C. F. Harding, appearing for Purdue, reported that programs giving pictures comparable to the ordinary newspaper print had been broadcast over distances as great as one thousand miles. He contended that this is the only kind of television likely ever to be available in ninety-five per cent of geographical America. It is not the best possible kind of service, technically speaking; but Mr. Harding reported that broadcasts of newsreels, showing horse races, men marching, and other events of the day were clearly received over long distances. Purdue's station is located at Lafayette, Indiana, yet its programs, Mr. Harding's evidence showed, were received by Fullerton, Pennsylvania, in such detail as to be called "photographic." After a considerable debate, the commission agreed to let Purdue continue its experiments but eliminated an operator in commercial television to allow more police radio.[5]

Just how many stations in the world operate on the medium high frequency range nobody can be sure. It accommodates a great number, as indicated by the guess that there are sixty-five thousand amateur stations operating around the world within that single frequency band. It also serves those wonder-workers who guide planes and battleships and perform other stunts of remote control.

The high frequency segment lies between 6 and 30 megacycles and its 1376 standard channels are world-wide in their service range. Here we must take into consideration another natural phenomenon which, like the Kennelly-Heaviside layer, conditions investment, bankable loans, and the prospects of entertainment. Sunspots, the Nilometers of the electronic age, recur every eleven years. Their vast time cycles of frequency have an important effect upon the pulsations racing in cycles of a hundred millionth of a second duration. Beginning in the medium high frequency division, each station must be allotted transmission bands sufficiently wide to permit shifts from one circuit to another with the changing phases of the sunspot cycle, which strongly conditions transmission paths, emission, power, directivity, and varied qualities of daytime and nighttime service.

By the time we rise to 6-30 megacycle zones, even wider bands must be allotted to cope with sunspot characteristics, magnetic storms, fading, and echoes off the Kennelly-Heaviside layer. Thus it becomes obvious that the number of users must be less, the farther down the scale. But the value for use increases, as the 6-30 megacycle frequencies are excellent in long distance communication.

Ships of the air and the sea, operators of coastal telegraphy, international broadcasting, mobile telephony and press service, fixed point-to-point telephony and telegraphy are characteristic users of this zone. Also present are the familiar government departments, the amateurs, and general experimental operatives—which latter class, for that matter, are salted all through. The important thing to note is that one finds here the same services struggling for a foothold which have clustered in every segment previously opened.

The commercial broadcaster with a good frequency down between 550 and 1600 kilocycles wants, like Mr. Heintz, nobody to shiver the spectrum in his area lest all shiver. But he wants more room, ever more room upstairs, for no man can tell what the inventors will loose upon the country next. The only defense against them is to grab everything available and yell for more.

8. Trouble in Heaven

ONE PORTION OF THE SPECTRUM IS VERY NEW. IT IS KNOWN as the ultra-high frequency segment, ranging from 30 megacycles ad infinitum, and it is the field upon which radio's titans are gathering for a tremendous struggle. In October, 1937, the Federal Communications Commission announced it would consider applications for licenses in the zone between 20 and 300 megacycles, and indicated its feeling in the matter with a little homily:

> The allocation of the ultra-high frequencies vitally affects several important broadcast services, namely: television, facsimile, relay, high-frequency and experimental broadcast services.
> The action taken by the Commission today with respect to television is merely one step of many which are required before television can become a reliable service to the public. Some of these many steps must be taken by the industry in the development of proper standards which in turn the Commission must approve before television can technically be of the greatest use to the public on any scale.
> Also the Commission, at the proper time in the future, must determine the policies which will govern the operation of television service in this country, particularly with reference to those matters which relate to the avoidance of monopolies. And the Commission must also in the future prescribe such rules and policies as will insure the utilization of

television stations in a manner conforming to the public interest, convenience and necessity, particularly that phase which will provide television transmission facilities as a medium of public self-expression by all creeds, classes, and social-economic schools of thought.

The investigations and determinations of the Commission justify the statement that there does not appear to be any immediate outlook for the recognition of television service on a commercial basis. The Commission believes that the general public is entitled to this information for its own protection. The Commission will inform the public from time to time with respect to further developments in television.[1]

After such a statement the commission can never plead ignorance of the issues. But what shall we make of its behavior? We know there is no question about the technical efficiency of television. Promoters and engineers are agreed that America leads in the technical aspects of all forms of communication by electricity. The real problem unquestionably is one of resolving conflicts between applicants for permission to perform.

Here is an example of the sort of mood in which these people approach the governing body:

I wouldn't have the intestinal fortitude—plain guts if you would rather have—even though representing the important police service I do, to stand before you in an attempt to confiscate the important band between 30 and 42 megacycles to the exclusion of commercial and other interests who have just need for such channels and for promoting the public good and welfare.

And if any service, governmental or otherwise, think they are going to get away with this without hearing from the police service which protects the lives and property of civilians in times of peace as well as in times of war, they are sadly mistaken. . . .

While a thug stands with drawn gun and cocked hammer we would betray a sacred public trust if we didn't seek our just share of frequencies and we are not going to be hoggish about it, either. . . .

We have no paid lobby but we do not intend to draw our punches for the benefit of the thug and to the detriment of the public at large.[2]

This is just Captain Donald S. Leonard serving notice, on behalf of the International Association of Police Chiefs, that these gentlemen, operating on the lower bands of the spectrum, want a place up where television may sprawl. He is indicating, rather melodramatically, the belief that police radio serves the public interest, necessity, and convenience sufficiently to warrant its continuance and expansion.

And William S. Paley, president of the Columbia Broadcasting System, holds that if private capital is going to continue doing the sort of broadcasting job it has started out to do in this country, its past investments must not be ignored.

I say this because there must be constant encouragement to capital flow if the people of America are to have the benefit of every technical discovery, every creative advance. For this reason, sudden, revolutionary twists and turns in our planning for the future must be avoided. Capital can adjust itself to orderly progress. It always does. But it retreats in the face of chaos.

We are on the threshold of a period of transition for the next couple of years. We should do everything in this period to advance experimentation. But we should do nothing to weaken the structure of aural broadcasting in the present band [of the spectrum] until experimentation in other bands has yielded to us new certainties.

For instance, allocations in the present broadcast band are such that even a few minor changes might upset the whole plan of the structure. The present layout is like a chess game. A single move can have almost infinite ramifications.

Probably the most important economic problem we must face—certainly the one uppermost in everybody's mind—lies in television.[3]

Not long after that declaration of his views, Mr. Paley made an extremely forehanded move in the interests of his company which is, on the whole, just a program service, with the duration of its life dependent upon the licenses of Columbia's outlet stations. On June 7, 1937, he filed with Securities and Exchange Commission at Washington an application for permission to sell shares to the general public in its going concern. The acceptance value may be judged from this: to initiate its chain of station outlets, the Columbia Broadcasting System expended in cash $1,600,000. The stock issued in June against this enterprise was sold to the investing public at market prices indicating a potential gross return of fifty-five million dollars upon the whole issue.[4] Mr. Paley has a reasonable right to assume that those investors will join him in an alert interest in any readjustments of the radio spectrum which might endanger their investment. If any of them had any fears concerning the six months' license provision, it was not recorded.

How does television imperil these vested interests of which Mr. Paley speaks so tenderly and Captain Leonard so vehemently? Here's an example: a "shadow," or unexplained interference, upset commercial and all other broadcasters along the Pacific Coast during 1935-36. For quite a while the scientists argued seriously whether or not the mysterious activity was not the long predicted message from Mars. Finally, it was discovered that the "message" was coming from diathermy machines with which doctors treat syphilis, arthritis, and give simple pleasure to hypochondriac movie stars.

V. Ford Graves, chief inspector of the Federal Communications Commission's western division, estimated that of the fifty thousand diathermy devices reported in use then by the American Medical Association, some forty-nine thousand were buzzing away in California. Now the most common (1935) model shoots heat into the human body by high frequency radio current of the 6-20 megacycle variety, but of relatively low volume. Newer types are rising both in power and in frequency, to threaten interference with radio activity in the entire upper area of the spectrum. No medical license was required either to make, own, or operate these increasingly popular instruments, as of 1937. Mr. Graves reported that one private citizen in Los Angeles, not a doctor, operated eighteen of them all day long to the interference of all forms of upper-band communication, even that of the United States Navy on maneuvers off San Diego.[5] One can imagine the howls that would arise if these fascinating titillators of the arthritic were limited by government fiat to use on rigid schedules. And yet, if the doctors are not restrained they may eventually blanket and scramble the radio communications systems in daily use, and send flickering distortions and shadows across the television screens of America to bring about a headlong collision of interests between the sick and well.

TROUBLE IN HEAVEN

But any possible trouble with doctors would be mild compared with the existing conflict outlined by Dr. C. B. Jolliffe, who resigned the post of chief engineer of the Federal Communications Commission to take a similar position with the Radio Corporation of America, one of the chief practitioners before the commission. He declares that the quality of a television picture is rigidly determined by the number of picture elements. The number of picture elements determines the frequency band which must be imposed on the radio frequency carrier. There is no short cut and no compromise. Consequently, we must face the fact that good television requires a wide band of frequencies. Good television can be included in a band width not less than 6 megacycles, but reduction in that band width will reduce the quality of the picture which it is possible to transmit.

When one considers the fact that all of the commercial auditory radio in the United States is jammed into an area on the spectrum between 550 and 1500 kilocycles, one can realize just how radio engineers and investment operators feel about the presence of "the great gobbler," television, in the zone just opened for licensing.

Dr. Jolliffe's doctrine is that reasonably good television can be broadcast over that area of the spectrum between 42 and 86 megacycles, but that each such broadcast must consume, vertically, 6 megacycles of spectrum space. But is that the whole case?

Say that a program is televised on a frequency of 42-48 kilocycles: would it then ripple across the continent to be picked up in San Francisco by really good receivers as easily as in New York? Not in the present state of the art, says Dr.

Jolliffe. The current "horizon" or perimeter of reception for visual broadcasts is about forty miles from the transmitting station. But the effects are not so limited, for interference and incoherent radio activity reach out on a radius of two hundred miles from the antenna.

Given the area between Boston and Washington to be served by television, Dr. Jolliffe works it out thus:

1. The distance between the two cities is, roughly, four hundred miles. Therefore, a station at Boston and one at Washington may emit programs on equal frequency, safe from effect upon each other's audience.

2. To send the same program from Boston to Washington by radio entails the use of "booster" stations, erected every forty miles to catch the program on one frequency and toss it on to the next station on a frequency of different register, which would interfere with no other station within a radius of two hundred miles. One quickly sees that the use of "booster" power involves complete exclusion of competition in the ordinary sense.

For if the booster receives on one frequency and emits on another, to escape interference within a two-hundred-mile range, and yet boosts the image only forty miles, the next station must likewise consume a third segment in order that its rebroadcast escape any interference. Each station carries over one frequency as it were, and picks up one new one. A single program sent from Boston to New York by booster power therefore consumes the whole series of television bands between 42 and 86 megacycles, and demands monopoly if Boston, New York, and way-stations are to enjoy the same program at the same time. Television, if it is to be successful, must approach universality of acceptance among the people. But must acceptance be at the price of its spectrum space given to a single operating concern? [6]

Dr. Jolliffe did not say so, but there are other means to disseminate television programs. A device owned by the American Telephone and Telegraph Company known as the "coaxial cable" is now in operation. By means of it, the television impulses can be propelled not forty, but as many hundred miles as one may wish between broadcasting stations. The broadcasters need only observe the law that stations within two hundred miles of each other broadcast on different frequencies. That is a state of affairs common to aural radio. And that is a state of affairs still involving monopoly. Not monopoly of the spectrum, it is true, but monopoly nonetheless; the sort of moral ascendancy the American Telephone and Telegraph Company now has over sound radio and sound motion pictures. It is necessary to understand the telephone company's interest in the radio spectrum if one is to appreciate matters of discussion further along.

In the zone between 1.6 and 30 megacycles, all of America's domestic telephony is laced by radio to ships at sea, the wire networks of more than sixty foreign countries, and airplanes in flight. As the spectrum exploitation jumps to 30 megacycles and above, the telephone system's interest jumps smartly along with it.

Lloyd Espenschied, radio transmission development director of the Bell Telephone Laboratories, Inc., states the prospects of his organization in this upper region to be of greatest importance. Two way service between ships, planes, and motor cars can be expanded and revised. Doctors can be called while driving in the country, can answer and exchange information. Armies in the field and navies on maneuver can function in closer contact with headquarters if they can shield their machines from interference and their messages from interruption.

In point-to-point service, the number of circuits that can be developed for simultaneous use can be vastly multiplied. Mr. Espenschied stated, in fact, that the upper megacycle radio channels and the Bell system's vitally important new coaxial cable have similar characteristics.[7] The cable will accommodate more than four hundred telephone conversations simultaneously, or one television program involving as many lines of definition. Thus we can see that the Bell system has great interest in the future of the radio spectrum above 30,000 kilocycles. Coaxial cables are priced at four thousand dollars a mile.[8] Radio channels cannot be estimated in terms of depreciation, upkeep, repair, nor, so far, of taxation. If a great network of cables is built, but not required for use in television, what becomes of investment and income?

Of course, the outlook for substitution is, as Mr. Espenschied is careful to claim, subject to current limitations of engineering powers. But, as he is equally generous to admit, there is no basis for assuming that engineering powers are even temporarily halted in the advancement of wide general uses of the upper megacycle zone.

The important thing to keep in mind is not that engineering is still pioneering in this area, but that engineering indicates a transmission band of 6 megacycles for the type of telephony Mr. Espenschied discusses and an equal band for television. It all works out very nicely from the engineering standpoint. But what about the telephone ratepayer?

Does he have to worry about the clearly implied conflict? Not, of course, if he has no complaints against the cost of service.

But the representatives of other, unsuspected, interests are not so casual. Geophysical prospectors who sound the inside of the earth for gold and oil and copper and iron by high-frequency current want to know the future of the radio spectrum. It means dollars to them. And it concerns the efficiency of government, too. Dr. J. H. Dellinger, of the International Bureau of Standards, demanded of the Communications Commission more than half the available frequencies between 20 and 192 megacycles on behalf of radio-using bureaus of the Federal Government. He got what he sought but not without limitations, for the agencies which keep them do so at the risk of being accused of that worst of crimes, "government competing with business"; and against the will and effort of many a person within as well as without the Federal Administration.[9]

Shall doctors treat syphilis and cancer to the detriment of naval communications? Shall television be set aside in the interest of field maneuvers of a tank corps in the Kansas prairies? It's everybody's problem.

The American Telephone and Telegraph system wants to expand, naturally. So (and it is no secret) does the Radio Corporation of America. Each sees the advantage of high-frequency transmission through the use of booster power stations as described by Dr. Jolliffe and admitted by Mr. Espenschied. And though they are agreed now not to fight, agreements have a way of fading before the necessity of self-preservation. Shall agreement in this case fade in the interest of wires or waves? The decision rests not with the contestants but with the referee, who has set about somewhat timidly to test his strength.

He has apportioned seven channels for television in the spectrum band between 44 and 108 megacycles, and twelve more between 156 and 300 megacycles, each channel 6 megacycles wide and providing for picture and synchronized sound. For old-fashioned sound radio, seventy-five channels are made available between 14.02 and 43.98 megacycles, and twenty-nine special police broadcast channels are provided between 20 and 40 kilocycles. Further provisions are made for aviation, geophysics, fixed point-to-point forestry, marine, and all the other familiar subdivisions of interest we met down at the lowest levels.

What stirs the blood of the radio man is the commission's announcement that applications for licenses must be filed with it before October, 1938, for allocation on a definite basis in 1939.

9. The Ethereal Klondike

RADIO STATION KMMJ, OF CLAY CENTER, NEBRASKA, BROADcasts on a frequency of 740 kilocycles with 1000 watts of power, by virtue of a "limited time" license from the Federal Communications Commission. That is, it must shut down at certain hours of the day to make room for some more powerful competitor. But KMMJ, if small, is far from humble. It conceives itself as something like the modern crossroads gossip, clad in straw hat and overalls, trotting from kitchen to kitchen with advice, news, and a sample case full of mail order sundries.

KMMJ eats breakfast, dinner and supper with Nebraska farmers and small town residents, not merely as entertainment but as a needed service. [That's how the station manager puts it.]
Fiddlers may stop fiddling and cowboys may cease yodeling, but the social and economic life of Nebraska-Kansas listeners is dependent upon KMMJ's news, weather and market reports and storm warnings.
Confidence, neighborliness, and friendly understanding are the keynote of KMMJ's effectiveness in producing sales. KMMJ for low cost results.[1]

There you have it. KMMJ is "the old trusty station," maybe, as it says it is. But in the end it produces sales.

There's nothing idle in radio's boast that it is a great business agent. Maybe the Communications Act neglects to admit formal recognition of the electron's merchandising power, but business does not. Dependent upon KMMJ and its kind are the sellers of soap, autos, candy, cough drops, poultry cures, lipstick, coal, coffee—a thousand staples of commerce listed in the sales manuals of the trade.

Monopoly may underlie it, but on the surface the radio business is a raucous, gaudy haggler's bazaar. Here's a fat merchant, pondering whether to risk his sales campaign on a moon-eyed comic and three crooning blondes. There's a station manager, looking at transmitting equipment and waving away a mountaineers' string band. Advertising agents, flashing smiles and figures with equal facility, poise themselves to tell even the most casual listener how they made beans, breakfast food, and chipped soap into national best sellers overnight and could do as much for his products.

"Flesh-peddlers" offer singers, dancers, dramatic actors, elocutionists and monologists capable of reciting poetry and feeble fable according to the tastes of a given community or patterned to go from coast to coast.

And everywhere men hold up watches, clocks, sundials—time, it's time they're selling. Fifteen minutes for nine hundred dollars. Twenty-one seconds for a twelve-dollar "station break." They have time for sale. The merchants and the clowns, the cracker barrel philosophers and the news commentators, the traders in equipment, talent, time, and tongues—all these stand a little in awe of the medicine men of radio, the fakirs who move through the bazaar wise in their privileges, aware of their authority. These are the lawyers and the engineers, dealers in the occult science of keeping peace between the hagglers and the bazaar masters, between the radio industry and the Federal Communications Commission.

And what is the radio industry? We know it really includes communication with ships at sea, world-circling telephony, educational and cultural nonprofit broadcasting, control of planes in flight, television, the operations of armies and navies, and that vast, fascinating playground of the amateurs. We know those things, but to the hagglers in the bazaar who give little enough thought to the fact that the bazaar belongs to someone else, and that the price of goods is controlled really by others who want things "stabilized in the interests of good economy," all of radio is commercial broadcasting. So let us say, for the moment, that the commercialized activity on the spectrum between 500 and 1600 kilocycles is the radio industry. What does it include?

First and foremost, there are the licenses of operation, and here is how the Communications Commission has reached a basis of allotting them. Sound radio frequencies require channels not less than ten kilocycles wide. This will just accommodate notes extending slightly beyond the range of the piano keyboard. The commission has subdivided the 500-1600 kilocycle portion of the spectrum into ninety-six channels of operation beginning at 550 kilocycles and moving in units of 10 kilocycles of width up to 1500 kilocycles. The 150 unassigned kilocycles of space serve as a guard against interferences from other phases of radio activity. Of the ninety-six channels, six have been allotted to Canadian radio, leaving ninety for the United States.

Failure to make specific provision for Mexico and Cuba

has led to considerable embarrassment and annoyance upon occasion. The most widely known incident is that of a Kansas City specialist in operations intended to rejuvenate. Barred from a license in the United States, the good doctor simply erected a powerful transmitter in Mexico, sent programs down from his Kansas City studio, and went merrily broadcasting on. In Havana thirty stations were erected, many directing programs to the United States [2]—an incidental problem of regulation neglected by the Communications Commission until Mexican and Cuban piracy of American channels became unbearable. Belated efforts at control have not yet accomplished a great deal.

But back to our ninety radio channels, so highly prized, so important to the bazaar full of yodeling cowboys, xylophonists, swing bands, and goods vendors. Of the total, forty have been set aside as "clear channels." By day, when interference limits the range of the electron, there may be as many as two stations in this "clear channel" division operating on equal frequencies and power but far enough apart, geographically, to prevent confusion. By night, however, one must shut down; and in all the geographic United States, forty broadcasting stations dominate the nation's homes. To operate on a clear channel, a station must have a power of not less than 5 kilowatts and, with one exception, not more than 50 kilowatts.

Radio Station WLW, of Cincinnati, however, has been singled out to try the principle of "super-power" broadcasting, and has enjoyed an exclusive privilege to operate on a 500 kilowatts power basis for more than three years. Last fall, as other station operators raised an intense outcry, WLW's owners saw fit to retain Charles Michelson, secretary of the Democratic National Committee, as their director of publicity; for in the face of complaint, members of the Communications Commission showed signs of restiveness toward continuation of the exclusive grant. So far WLW continues its "experiments" free of competition.

The matter of assigning clear channels, obviously, was of basic importance to the whole radio structure when the broadcast spectrum was rebuilt by the Federal Radio Commission after 1927. Rural and remote areas cannot be served except by the high power, clear channel stations. Obviously, they ought to have been assigned on a geographical basis in order to assure the greatest possible facility of reception for every citizen, no matter where residing. Whether that perfect state has been attained is a matter of continuous and acrimonious debate among authorities which, if not convincing in any other respect, does give evidence of the necessity for critical interest by the public in the assignments of television licenses hereafter to prevent any basis for charges of inept distribution of service.

Another important question concerning the assignments of clear channels concerns the kinds of licenses granted these dominating outlets of entertainment and trade stimulation. It is a curious fact that not one of the clear channels is dedicated exclusively to cultural or educational pursuits. Every one is in the hands of the commercial operator, whose primary interest naturally lies first with his pocketbook and only as occasion demands with the public interest, necessity, or convenience. It is not true that only commercial operators are able financially to maintain clear channel radio stations. Colleges, trusts, endowments for educational uses, privately operated philanthropical institutions, and municipalities stand ready to serve the nation by radio, if only they can obtain licenses for operation.

A most significant fact about the social possibilities of the radio industry is that thirty-eight broadcasting stations persist in operating on the domestic spectrum of the United States without profit in spite of all the difficulties they endure in the way of inadequate frequency allotments. We insert (see Appendix B) the full list of nonprofit licensees, with call letters and locations of stations, as a matter of historic interest, but omit the power and frequency ranges of those stations, since they are generally so inferior that the programs are unavailable outside regional or community areas.

Students of institutional propaganda will be interested to note that only one state and one city operate radio stations directly on a public service basis.

The high percentage of state university license holders in the midwestern section of the country is a curious fact, for which we can offer no patent explanation. It is probable that radio's valuable uses in connection with farm operations (weather reports, market news, and education on growing of crops) were the moving factors.

The noncommercial broadcasters have made intense efforts to improve their radio status, but without success. In 1934, shortly after passage of the Federal Communications Act, a conference of all parties interested in a proposal to fix definite percentages of the total available radio frequencies for noncommercial broadcasting was called in Washington by the Communications Commission.[3] (Perhaps it ought to be explained that the word "noncommercial" is used simply to make a distinction between stations operated for money profits and those maintained primarily for cultural or propaganda purposes.) At the 1934 conference, such representative organizations as the National Educational Association, National Catholic Educational Association, National Association of State Universities, International Council of Religious Education, Children's Bureau of the Department of Labor, and National Association of Broadcasters entered statements of position. In all, one hundred and thirty-five witnesses appeared; and they filled fourteen thousand pages with testimony.

The commercial broadcasting industry, from manufacturers to station licensees, presented a united front of opposition. They claimed a plant investment in that year of $25,041,327 which they said was jeopardized by the threat of endowed, nonprofit radio; and they claimed that they were then allowing about twelve per cent of their own expensive time to go for educational, nonprofitmaking programs.

This, of course, is not convincing. Definite profits, in the form of prestige and cultural standing, accrue to the stations from such programs. In addition, they save considerable sums of money: distinguished talent is acquired free of charge to fill in time which the studio otherwise would have to occupy at its own expense with "sustaining programs," since the Communications Commission prohibits the shutting down of transmitters during allowed broadcasting periods.

The Federal Communications Commission finally recommended that no fixed percentages of radio broadcast facilities be allocated by statute to particular types or kinds of nonprofit programs, or to persons identified with particular types or kinds of nonprofit activities.[4]

It stated that the present law is flexible enough to allow such allocations if they are warranted and of this there can be no doubt, in view of the mandate to grant licenses only on the basis of public interest, necessity, or convenience. The only difficulty is that to allow a nonprofit applicant a frequency in the average community, the commission is put to the painful task of throwing some currently existing commercial operator off the air. It appears not to have the will to do that with ease. Conversely, it orders the noncommercial broadcaster to defend himself against a commercial applicant for the frequency already in use by the noncommercial operator.

In one notable case, that of WNYC, the municipally owned station of New York City, a commercial operator was able to convince the Federal Radio Commission that he possessed powers to serve the public interest, necessity, and convenience, superior to the richest and largest city on the North American Continent. As a result, a first rank frequency was taken from New York City's municipal government and given to the "business man," and the city finally was awarded a very inferior substitute frequency.

The WNYC affair occurred before the days of the Federal Communications Commission, but that body has not shown the slightest disposition to restore the superior frequency to WNYC. Thad H. Brown and Eugene Sykes, who were carried over in the new organization from the old commissions, have never been called on by Congress even to explain the basis for the action, but Brown's position as a commissioner was challenged very sternly by New York City, at the time of the hearing on the case by the old Federal Radio Commission, on the ground that before becoming a commissioner he had acted as counsel in examining the case for the commission, and "on the further ground that at certain times he had been brought in person into this hearing by representatives of station WMCA and WOCH and the hostile attitude of his representations reflected his attitude in connection with the application of WNYC—." An effort was made to have the other members of the commission remove Brown from office during the consideration of the case because of the charges against him, but he was successful in maintaining his status.[5]

The Communications Commission, in recommending no fixed percentages of radio time for services for nonprofit activities, held that no feasible plan for a definite allocation of broadcast facilities to nonprofit organizations has been presented. The hearings, it was claimed, developed no evidence of a real demand on the part of the great body of nonprofit organizations or on the part of the general public for the proposed allocation of definite percentages of broadcast facilities to particular types or kinds of nonprofit activities.

It would appear that the interests of the nonprofit organizations may be better served by the use of existing facilities, thus giving them access to costly and efficient equipment, and to established audiences, than by the establishment of new stations for their peculiar needs.[6]

That is a most reasonable analysis of the facts, and it appeals for agreement, unquestionably. There is just one flaw. Nothing is stipulated to insure that the nonprofit broadcaster can get time to operate on these costly and efficient devices when he thinks he should, or when the public interest, necessity, or convenience patently dictate.

Nothing, moreover, is stipulated to insure that what he has to say will be protected against censorship or emendation by the station licensee.

The commission, as an antidote to its negative findings, announced the appointment of a Federal Radio Education Committee, designed to "eliminate controversy and misunderstanding between groups of educators, and between the [radio] industry and educators," and to "promote actual cooperative arrangements between educators and broadcasters on national, regional and local bases." [7]

This committee was established on a grandiose plane. The United States Commissioner of Education was made its chairman. Such distinguished prelates as Dr. S. Parkes Cadman and Father G. W. Johnson of the Catholic University of America found places on it. William Green, president of the American Federation of Labor, and Dr. Robert A. Millikan, of the California Institute of Technology, gave it breadth of connections.

Parliamentarians will find it interesting that of the thirty-nine members of this distinguished committee, a tight, solid minority of eighteen were openly identified as representatives of commercial broadcasting, which had already put itself on record as completely opposed to the theory of fixed percentages of radio time for noncommercial purposes. In an organization of thirty-nine members, a united bloc of eighteen votes generally allows no freedom of action for the remainder on controversial issues.

At any rate, the record clearly shows that the Communications Commission's committee has been unable to insure that an educational, nonprofit broadcast can be made when, as, and if the judgment of the would-be maker dictates.*

Education, whether in classroom or studio, is supposed to proceed on classic lines of free speech and free thought. Will any studio manager come forward to demonstrate that such is the case when the speaker threatens to jeopardize the "listener-appeal" of his own carefully cultivated audience? There will be further delving into the record of this free speech issue a little further along, after identification of the remaining types of frequency allocations.

The next best grade below the clear channel is the high power regional channel. There are four of them serving special areas. They operate generally upon 5 kilowatts of power. For coverage of large cities and their suburbs, there are forty regional channels, operated on 250 watts to 5 kilowatts of power. From three to seven stations are assigned to each channel, but upon such geographical distribution that interference is not serious. There are six remaining local channels, of 100 to 250 watts in power, accommodating approximately fifty stations each. These serve small cities and towns, and fit in between the important points on the dial in the big city.

The total number of stations licensed to operate on the domestic spectrum currently is seven hundred and four, of which one hundred are sharing time on clear channels, nine on high power regional, two hundred and seventy-four on

* Its chief accomplishment has been the establishment of an exchange for program scripts which, in the 1937 fiscal year, furnished 108 stations in 41 states with a total of 966 programs, according to the report of the Federal Communications Commission for that period (page 50). More than 1700 local groups used this service, and received sixty thousand copies of scripts, manuals, and glossaries of radio terms.

regional, three hundred and seventeen on local, and four on special service channels.[8] Of course, not all can operate at a given time for many have to shut off at sundown when changed atmospheric conditions allow wilder gyrations and bouncing of radio waves off the Kennelly-Heaviside layer to threaten interference all over the spectrum. A little known but very important subdivision of American radio is that offering international broadcasts not subject to reception technically on the average home set here. There are thirty stations licensed to operate in this field. Of these, seven have been granted to the Westinghouse Electric and Manufacturing Company, five to the Columbia Broadcasting System, four to the National Broadcasting Company, four to the Worldwide Broadcasting Corporation (an organization operating under the auspices of Harvard University), three to the Chicago Federation of Labor, two to the General Electric Company, two to the Crosley Radio Corporation of Cincinnati, two to WCAU Broadcasting Corporation of Newton Square, Pennsylvania, and one to the Isle of Dreams Broadcasting Corporation of Miami Beach, Florida.[9]

International broadcasting is a new phase of American radio, as yet not wholly developed in matters of policy. As a rule, directly sponsored broadcasts on a commercial basis are not allowed, but nothing prohibits the international broadcaster from rebroadcasting a commercial program of the domestic American type. This sort of thing is considered extremely useful for promoting trade in Latin America, where European propaganda programs of highly charged political and commercial content have been growing in popularity.

THE ETHEREAL KLONDIKE   93

There must be, within the next few months, some clear declaration of government policy on this international program service if the United States is to stay clear of diplomatic encounters; for commercial broadcasters, as the upper frequency portion of the spectrum is made available, are going to seek to offer their programs to the world.

Radio may be an industry characterized by an erratic behavior of finance but it suffers no lack of investors. They appear to be eternally fascinated by its statistics. There are more radio sets than telephones installed in American homes. Approximately thirty million instruments capable of being tuned in on the broadcasting bands are operating in residences, automobiles, and boats. They have an estimated investment value of $3,000,000,000, and ninety per cent of them are always in working order.[10] In 1935 listeners spent $150,000,000 just for power to operate these sets,[11] and sat by their radios approximately one billion man hours a week.[12]

There can be no doubt that the business men of America believe in the value of radio advertising. In 1936 they spent in excess of $115,000,000 just for radio time space.[13] How much more they paid the artists and musicians who performed for them is not known but the sum must have been considerable, in view of the commonly publicized salaries of thousands of dollars paid star performers for single weekly performances of less than one hour. The weekly payrolls of five hundred and fifty-seven stations reporting to the Department of Commerce in 1935 showed salaries totaling $428,401 paid to 13,139 employees.[14]

Stations are valuable as properties, too, in spite of the six months' license terms always threatening "governmental interference." Perhaps the most striking example of confidence in the future of radio is presented in the case of station KNX, of Los Angeles, broadcasting from a physical plant valued at $217,237.85. KNX claimed a value of $236,520.21 on its stock at the time it transferred its license to the Columbia Broadcasting System, and showed net earnings of $107,933.70 for a twelve month period.[15]

Yet, at the moment when its management had been accused by an investigator for the Federal Communications Commission of more than forty violations of regulations, KNX succeeded in transferring its license to Columbia for a consideration of $1,250,000, and the investigator actually found himself transferred to a position from which he could not push the case for disposition.[16]

Broadcasting stimulates many collateral business activities. The investment in transmitters and receiving sets, the expenditures for electric power, maintenance, and repair of equipment, the impetus to sales of advertised products with the subsequent increment in employment and money turnover in the affected industries—these are just a few. What will be the effect of television upon general business?

As to the quality of a program and acceptance by the audience, the radio industrialists have no certain test except sales of products advertised in conjunction with it. If a program results in larger orders, it is a good program. You may draw your own conclusion as to what is appealing from the fact that 25.9 per cent of the radio-time dollar in 1936 bought attention for programs advertising food and beverage products. The next largest segment, 21.6 per cent, was invested in promoting drugs and toiletry sales, and the third largest, 15.4 per cent, was spent by the automobile companies.[17]

What is the advertiser to do? Edgar Bergen and his

dummy, Charlie McCarthy, on the Chase and Sanborn coffee program, are sheerest nonsense, yet every test shows this feature the most popular in the country. Therefore, Charlie McCarthy sets the pace for the advertisers of food and beverage products.

A quondam medical student, M. Sayle Taylor, was so successful in 1936 as "The Voice of Experience," broadcasting solutions of love problems while advertising hair tonics and itch salves, that he even started a popular magazine, and received mail literally by the truckload. His type of personal advice, the dramatic love story and the beauty hint, have come to set the standard in drugs and toiletry programs.

The enormously popular Major Bowes' Amateur Hour, developed with Chase and Sanborn as a food selling feature, has been tried out by the Chrysler automobile concern, but more typical of automobile sales promotion by radio are the Ford Sunday Evening Hour and the General Motors Concert. These two offer entertainment of excellent quality from the artistic standpoint. The advertising is unobtrusive and brief, to the point and soothing. It is accepted as a reasonable bargain for good entertainment. Yet what will become of this sort of program when television comes? There is considerable doubt that advertising will be successful when presented to the eye as well as the ear. Certainly the motion picture industry has never succeeded in finding a spot for advertising in its programs—and it has tried, strenuously.

The clowns and the fakirs, the engineers and the violinists of this seething radio bazaar have a lot of readjusting ahead of them in the next five years. They must choose a side and get on it, if they are going to stay in the public favor. Superficially, it may appear that radio is in a ferment of competition and likely to remain that way as television comes. But stubborn facts proclaim that ninety-three per cent of all the allotted power in radio broadcasting is assigned to stations under the aegis of just three concerns, the Radio Corporation of America, the Columbia Broadcasting System, and the Mutual Broadcasting System,[18] and that in television there will not be room for all three in a single city unless some drastic action is taken by the Federal Government.

KMMJ and its homilies about breakfast and weather may be eliminated any time in the interests of scientific or commercial progress, and only the farmers and small townspeople of a remote area in Kansas and Nebraska would complain. But dissolve the National Broadcasting Company on the same basis—and then what? The NBC's parent, the Radio Corporation of America, has been forehanded. It is preparing for the great encounter—and has been preparing—for a long time. Just what is its status concerning television? We shall see presently.

10. Microphones and Censors

THERE IS HARDLY A CORNER OF THE GLOBE TO WHICH NO radio broadcast penetrates. And, contrary to common belief, most countries allow commercialized broadcasting. Great Britain, Germany, Russia, Japan, and Italy have developed radio to high standards of technical proficiency, but hold it exclusively the government's property. In the case of Great Britain, a public corporation known as the British Broadcasting Corporation conducts programs in accordance with the terms of a Crown charter. The cost of operation is paid by taxation of receiving sets at the rate of ten shillings a year.

But while there are commercial stations in Europe, by far the greater volume of broadcasting power is consumed to broadcast noncommercial programs. That is the case, generally, around the world. In South America, where broadcasting is fairly new, no clearly defined continental policy exists, but radio is very popular. Brazil, for example, reports about three hundred and thirty thousand tax-free receiving sets among its three million population, who are offered programs by sixty commercial and government stations.[1] Chile reports fifty such stations, broadcasting to sixty thousand tax-free sets distributed among the four million three hundred thousand people.[2] Other South and Central American nations show about the same statistical relationship between population, stations, and sets.

In Africa, the French possessions of Algeria and Morocco allow commercial broadcasting, with one such station in each territory. There are about forty-two thousand taxed receiving sets in Algeria, and twenty-nine thousand in Morocco for the respective populations of six million five hundred thousand and five million.[3]

The Australian Commonwealth has about seventy-five commercial broadcasting stations, serving eight hundred and fifty-five thousand taxed receiving sets among the six million six hundred and seventy-seven thousand of population.[4]

Canada reports in excess of one million six hundred thousand radio sets, served by seventy stations, operated variously by private licenses, the Canadian Broadcasting Corporation, townships, and telephone companies. Commercialism is allowed under the general, but not ironclad, system of regulation imposed by the CBC upon its own station outlet. Receiving sets are all taxed.[5]

In Asia radio is making peculiar progress. Of course, in Japan it is operated on strictly governmental terms, and the military dictatorship exercises the most rigid sort of censorship and propaganda dissemination. In China radio is literally the only means of national instantaneous communication. The extent of radio activity in China is unknown to the outside world, but at least three hundred thousand tax-free sets are supposed to be floating around in the proclaimed Chinese Republic. At Shanghai alone there were thirty-seven stations [6] (more than in any other city in the world) broadcasting up to the time of the Japanese invasion of 1937. The state of broadcasting at Shanghai might be said to exemplify the classical state of life in China: greater density of propagation than anywhere else in the world, and chaos rather than orderly procedure.

Many Americans have condemned commercialized broadcasting, and would like to remake our system upon the British model. In 1936, a special committee, under the chairmanship of Lord Ullswater, made a careful study of the British Broadcasting Corporation to determine whether to renew its charter, and reported that its programs had widespread approval among the British public.[7]

The committee recommended a ten-year extension of life for the BBC, with provision that a board of seven governors should be selected by the Crown. It was stipulated that the governors should not be specialists or representatives of any particular interests or localities, and that "the outlook of the younger generation" should be reflected in some of the appointments. Minor issues, measures of domestic policy, and matters of day-to-day management should be left to the free judgment of the Corporation, the report held.

The Minister responsible for broad questions of policy and culture would be a selected Cabinet Minister in the House of Commons, free from heavy departmental responsibilities and preferably a senior member of the government. This Minister should have the right of veto over programs and the duty of defending the broadcasting estimates in Parliament, but technical control should remain with the Postmaster-General. The BBC should have the right to state when it is broadcasting an announcement at the request of a governmental department, and the right of direct government control in case of national emergency should be maintained.

As to handling of political and topical matters, it was resolved:

That, continuing present practice, the B.B.C. should refrain from broadcasting its own opinions on current affairs;
That the broadcast news service should be unbiased and dispassionate; that the B.B.C. should have a free choice as to the sources and methods of obtaining news . . . that controversial broadcasts should continue, discretion remaining in the hands of the B.B.C.;
That attention should be directed towards Parliament as the natural center of political interest, that Parliamentary news should hold its place in news bulletins and that, if broadcasts by a Parliamentary observer are continued, the observer should be provided with adequate facilities; that the B.B.C. should regularly consult the Parliamentary parties on major political issues; that during a general election campaign the time available for political speeches should be allotted by agreement between the parties [but what to do in the event the parties disagree is not stated] and that all political broadcasting shall cease three days before the poll. . . .
That direct advertisement should remain excluded from the broadcast service; that "sponsored" items need not be entirely excluded, especially in the earliest stages of television broadcasting, but that their admission should be carefully regulated by the B.B.C.; that the responsible departments should take all the steps which are within their power with a view to preventing the broadcasting from foreign stations of advertisement programs, intended for this country, to which objection has been taken.

These recommendations form, substantially, the principle of British broadcasting. As to technical standards, the committee urged "that the BBC and the wireless trade should jointly examine the possibility of designing and putting on sale at a low fixed price a standard receiving set."

The government had already ordered a pooling of patents by television inventors in order to develop national standards in that field, and the policy committee recommended considerable investment by BBC in television. All these recommendations have been acted upon favorably.

In comparison with the Communications Act of 1934, this British policy appears restrictive in the extreme. The broadcaster in the United States is free, technically, so long as he commits no libel, slanders nobody, permits no obscenity or profanity. He need only be sure of his license renewal every six months. The British concern must work in close harmony with the government at all times, and observe caution in all things.

The American politician can buy as much broadcasting time as he pleases, and the station manager is his own judge of how much free time he wants to donate to any campaigner. The only stricture is that every bona fide candidate have equal opportunity to buy time. But in Britain equality of opportunity rests upon some nebulous "agreement," and all radio campaigning must end three days before the voting begins.

Britain permits some discreet violations of its prohibition upon advertising, and concentrates upon uplift and education in programs and the raising of standards of technical performance. In the United States, while transmitters are strictly regulated, specifications for receiving sets are ignored. This is a condition which we know cannot obtain in television. Whether it likes to or not the Federal Communications Commission must prescribe set qualities.

The British appear uniformly pleased with their compromise between freedom and regulation. No such state of affairs obtains in Germany, Italy, or Russia. In these three countries radio is dedicated to whatever the administration desires. And while other European nations allow some commercial stations, as in the cases of France, Poland, and Rumania, vastly the greater portion of European radio is government owned or dominated. The nature of the programs is summarized by one of the best-qualified critics in the United States, David Sarnoff, as follows:

> I have listened, at many different times, to programs originating in every country in Europe. They have given me a great deal of excellent music. But many of them have also given me statements glorifying or condemning political and economic philosophies, creeds and personages in terms which could not conceivably be employed on the air in the United States.
> They have presented as news, statements contrary to fact or discolored by partisanship; they have omitted from what purported to be news, facts of essential importance. By any definition, a good deal of this broadcasting is propaganda, and some of it highly objectionable propaganda. . . .
> The diligent short wave listener who dials around the European stations and hears conflicting and contradictory interpretations of world news and politics soon develops a healthy propaganda immunity.[8]

He can develop no such immunity, however, if his loudspeaker is hooked to a wire cable, as, it is reported, may soon be the case in Germany, nor can he spin the dial to every station if standards are not sufficiently high to give him sets adequate to the task.

The critic of censorship and propaganda in European

radio says nothing about the parallel cases in America, but the facts argue that he should if he wants to be fair. Owen D. Young, one of the founders of American radio, discussing radio and censorship at Rollins College, said:

> Freedom of speech for the man whose voice can be heard a few hundred feet is one thing. Freedom of speech for the man whose voice can be heard around the world is another. ... The freedom of speech now depends upon the exercise of a wise discretion by him who undertakes to speak. . . .[9]

In other words, you are free to speak, so long as you speak wisely and with discretion. Freedom is not absolute, but dependent upon convenient circumstances. The American Civil Liberties Union regards Mr. Young as somewhat sinister:

> Mr. Young not only admits that free speech on the radio labors under special restrictions, but practically threatens that speakers who do not exercise what Mr. Young and the National Broadcasting Company consider to be a "wise discretion" will suffer the fate of the Republican Party, Hamilton Fish and Norman Thomas, and find their highest ideals displaced in favor of an advertisement for laxatives.

Those are strong words. It is difficult to agree that Mr. Young goes so far as overtly or covertly to threaten anybody. Indeed, one is forced to admit that he has come close to stating the real conflict between established social institutions and dynamic technological developments. But just what does happen to freedom of speech on the American radio?

Right at the broadcasting station, the censor, in the person of the station manager, begins to work by refusing to sell time or fulfill a contract for performance if he disagrees with the viewpoint expressed in the proposed program, even though it violates no law. The station censor always demands written copies of speeches in advance. And while the Communications Act gives him no power to edit the legal or political views of a candidate for office, who can say what may go into the making of the original contract for purchased time? The censor can and does drown out or shut off a speaker in the middle of a broadcast if the written and approved text is departed from; or he may set the program for an hour late at night when the audience is at a minimum.

But who actually and directly censors radio in the United States? Henry Adams Bellows, a former member of the Federal Radio Commission, and former vice-president of the Columbia Broadcasting System, says: "The only possible answer to the question, 'Is radio censored?' is an unqualified 'yes.' It is censored by the Federal Communciations Commission." [10]

The station operator always knows he is going to come back to headquarters in six months for a license renewal, while the complainant against censorship is only an individual who has transferred discretion and control to the government agency.

Mr. Bellows did not speak solely on the basis of abstract belief. In 1933, a group opposing recognition of Soviet Russia sought to buy time from the Columbia Broadcasting System. Walter L. Reynolds, secretary of the organization, reported that Bellows refused to make a contract, and frankly stated that no broadcast that was in any way critical of any policy of the Administration would be permitted over the CBS; that the Columbia System was at the disposal of President Roosevelt and that they would permit no broadcast that did not first have his approval.¹¹ The only commitment that could be got from Bellows was that if Reynolds would get permission from the President or from the Secretary of State, in writing, that they would have no objection to such a program, he would give the matter further consideration.*

Columbia Broadcasting System is not the only organization timid of criticism toward the Administration. W. E. Myers, New England representative of the National Broadcasting Company, in a letter to the American Legion, pointed out that criticism of the Economy Act of 1933 must stop.

> The American Legion, in its patriotic support of the United States Government, has always had, and shall always continue to have, the privilege of presenting its views over these stations.
>
> But we are obliged to impose regulatory and prohibitory "rules of the game." These are prescribed by our editorial policy, customary among all broadcasting stations, and have their origin in regulation of the Federal Radio Commission.
>
> Particularly at a time of national crisis, we believe that any utterance on the radio that tends to disturb the public confidence in its President is a disservice to the people themselves, and hence is inimical to the national welfare.¹²

The "disservice" of the American Legion was to criticize the President for leading Congress to pass the Economy

---

* Bellows' attitude appears extreme. Criticism of the Administration has occurred and continues to occur regularly. Commentators like Boake Carter, George Sokolsky, Dorothy Thompson, and William Hard attack Roosevelt policies without known interference.

Act, the result of which was to reduce benefits to veterans of American wars. Station WIRE, of Indianapolis, refused to accept an NBC broadcast by Earl Browder, candidate of the Communist party for President in 1936, on the ground that Indiana law bars political parties urging the overthrow of the Government by force or violence. On the CBS network, Browder and Hamilton Fish, of the Republican party, undertook to debate the future of American politics. The fourteen stations of the New England Yankee Network flatly refused to carry Browder, regardless of the policy expressed in the Communications Act. Instead, it broadcast dance music. But when Fish came on, his reply to Browder went over the Yankee Network unrestricted.[13]

However, the Communist party was by no means the only victim of censorship and restriction in the 1936 campaign for the Presidency. That same Hamilton Fish, undertaking to open the Republican campaign for the Presidency on station WHN, found his contract canceled, and was led to say: "People have been talking about radio censorship, but this is the first time we have a definite case."[14]

In January, 1936, the Republican Party Campaign Committee presented a series of skits on the "American way of life" and what the New Deal was supposed to be doing to it, only to find that neither network would broadcast them. David Lawrence, a leading commentator for the party in newspaper columns, complained that the radio, it appeared to him, was available for the sale of laxatives but not to sell ideas that "relate to the very foundations of American Government."[15]

The Grand Old Party was to suffer even more painful indignities before the campaign was ended. On one occasion

Senator Vandenberg, of Michigan, undertook to conduct a "debate" with the President of the United States by playing off transcriptions of actual broadcasts by Mr. Roosevelt and contrasting them with later Presidential utterances. He actually got on the air with this novel presentation, but within a few minutes found himself shut off on many stations, interrupted at others, and altogether the center of a studio tempest.

Political parties are not the only sufferers from censorship. News commentators must guard themselves carefully. Many stations prohibit or carefully suppress news concerning strikes and labor problems, religion, health, and plain gossip. Of course, there is a valid basis of editing in the interests of good taste and relative value of incidents. The radio is an unguarded instrument in the American home. It was heartily welcomed and people are attracted to it. There probably was sound ground, in the fall of 1937, for refusing to allow a broadcast on venereal disease by Hugh Johnson, ex-general of the army and erstwhile co-ordinator of the National Recovery Administration, who is uncommonly free spoken as a newspaper writer.*

But when Dr. Thomas Parran, Jr., now Surgeon-General of the United States, undertook to speak on public health needs in November, 1934, he found that the censorial pen would destroy the point of his message; and he refused to make the broadcast. Here is what Dr. Parran wanted to say, but was forbidden to:

We have made no progress against syphilis, though its

* The text of General Johnson's quashed address was published in newspapers, it ought fairly to be added, and no criticism appears to have developed.

results crowd our jails, our poorhouses, and our insane asylums. Yet there are specific methods of controlling it, better known to science than the methods of controlling tuberculosis. We need only to do what we know how to do, in order to wipe out syphilis. . . .

In my philosophy, the greatest need for action is where the greatest saving of life can be made.

I consider, then, that our greatest needs in public health are, first, the levelling-up of present services so that every community may receive the benefits that have long accrued to the leaders; and, second, a frontal attack by all communities against maternal mortality and deaths among new-born infants; against dental defects and faulty nutrition; against tuberculosis, where splendid gains have been made; against cancer and syphilis, where we have done little or nothing.[16]

Had radio allowed Dr. Parran to make that address, it would have acted in its own best interests to prove itself a valid agency for the social good, for within a year Dr. Parran became Surgeon-General of the United States Public Health Service and instituted a campaign of public education on the detection and cure of venereal diseases which has since received nationwide approval.

Some of radio's censorship is genuinely comic. On one occasion, according to the Civil Liberties Union, a gentleman undertook to strike a blow by way of the CBS station in New York City on behalf of worms as the proper bait for trout fishing. It appeared that the worm-lover, one Fred B. Shaw, was considered a sinister radical and maverick by fellow members of the Izaak Walton Club because of his predilection, and that his only colleague in the art of catching trout with worms was Calvin Coolidge. The Izaak Walton Club complained to Columbia that it did not want to be identified through Mr. Shaw with what might be considered either advocacy of worms or Coolidge.

"Mr. Shaw refused to back down on the question of worms, and CBS refused to allow worm propaganda to endanger the Nation." [17]

Major General Smedley D. Butler, retired, of the U. S. Marine Corps, has always had a reputation for language on a par with that of the other military gentleman, General Johnson. General Butler acquired newspaper publicity for quite a while by getting himself cut off the air. At one time, during an address to the National Convention of the Veterans of Foreign Wars, he undertook to criticize the policy of the Agricultural Adjustment Administration concerning its hog-killing program, and was cut squarely off. Finally, it was reported, he worked out a sort of cursing code with the companies by means of which he was reportedly authorized by NBC and CBS to use three "damns" and two "hells" every ten minutes. The General, apparently tongue-tied by such tactics, despairingly told an audience: "I can't talk soldier's language before these deodorizers [the microphones], so prepare yourselves for seventeen minutes of tripe and bedtime stories." [18]

The care with which Walter Winchell's broadcasts are prepared is widely discussed within the radio trade. Winchell gives his audience the impression of rip-snorting, ad lib chatter, but the fact is that he must follow an approved script. Until he learned to conform, he suffered badly from the censor. For example:

I think the best joke about New York Supreme Court Justice Crater, who is hiding about ten blocks from here, was the one told by Detective Elinson. He says that Crater probably got lost in one of those robes the judges wear.[19]

However innocuous and pointless that anecdote may appear to you, it sounded libelous and indiscreet to the radio censor. Winchell works around such restrictions nowadays. So do others. Norman Thomas, leader of the Socialist party, has discovered how to express his views without oppressive cutting and alteration of text, he reports.

Father Charles E. Coughlin, when he began to build up his radio audience in 1930, was violent in assault upon financiers and Communists, to the embarrassment of the Columbia Broadcasting System, which warned him that unless he allowed censorship he would be cut off the air. At the next opportunity, Father Coughlin asked his unseen, but supposedly palpitating, audience if it wished his attacks on the money changers and the un-American radicals to be curbed by the radio. The flood of mail that came pouring in settled that question in short order.[20]

But not every broadcaster is constantly threatened with censorship.

11. Ethics and the Listener

RADIO, LIKE THE PRESS, HAS A TENDENCY TO FLY INTO A RAGE whenever accused of pandering to low tastes and submitting to pressure from advertisers.

The press, on the whole, is fairly free from the sort of direct subservience to the government that characterizes radio. It has a guarantee in the Constitution against overt abuses, and needs only to guard itself against such indirect threats as revocation of second-class mailing privileges granted by the Post Office Department, and governmental tinkering with taxes, labor matters, and violation of the lottery laws. Of course, the publisher and the manager of the department store find themselves quite naturally on the same side of a crucial issue many times, but even so, newspapers as a rule do not ignore in toto the news concerning major events and shocking social abuses. Radio, the evidence shows, quite commonly dodges these distasteful duties or presents dehydrated versions. The publisher may sell space in his newspaper to a business man and still attack him in the news pages and editorial pages. He does not often do so, it is true, but he can—and, let it be added, he has done so. Furthermore, the press does print news critical of itself if those strictures are uttered by persons holding public attention.

The case of Mr. Justice Black of the Supreme Court, and numerous instances involving President Roosevelt, demonstrate this. Mr. Black, on his return from Europe, refused to discuss his membership in the Ku Klux Klan with newspaper reporters on the ground that their publications might refuse to carry his statement in full, or might distort or criticize it. Instead, he said, he would go directly to the people by way of the radio. And so he did, to a people who had been informed by the newspapers of his refusal and his intentions. Mr. Roosevelt, in a radio address to the nation shortly after he lost his campaign in 1937 to alter the status of the Supreme Court, pointedly omitted the newspapers from those media which he said had been doing noble service in educating people concerning the functions of government. He did, however, praise the radio and the motion picture. His attitude was not suppressed by the newspapers, but was made the subject of considerable editorial criticism and analysis.

Readers of newspapers have come to expect a fairly comprehensive report on the state of things in the world, and, by turning to others in which they have more confidence if such are available, or by taking interest in none if the press of a city is uniformly unreliable, they censure publications which persist in abusing the public credulity. They are not disposed as a class, however, to cancel their subscriptions and refuse to have anything more to do with a publication which presents them with one or two offensive stories out of many in the day's events. Whenever an editor or publisher wearies them with persistent campaigning on some political issue, they either find other newspapers to read or ignore his tirades and enjoy the features of his paper which they really like; and no editorial campaign which lacks merit and public interest succeeds or can even be maintained moderately well for very long.

The radio station manager is in no way comparable to the editor of the newspaper. It is the editor's task to sell newspapers in the tradition of informing, entertaining, and educating with a balanced presentation of news, editorial opinion, and features, such as comic cartoons, serial stories, and crossword puzzles. The station manager, on the other hand, sets out first to induce as many people as possible to tune in on his frequency, and second to sell their whole attention to the highest bidder. In the case of the newspaper, the medium is flexible and capable of expansion. After the editor has caught the public's attention, he can open his columns to advertisers. And with the income from advertisements he can open his news and feature columns still wider. But the station manager's problem is considerably different.

He has a rigidly limited total of time-space. The more advertising time he sells, the less he has of what might be called editorial content in his day's presentation. Time which is not sponsored by some advertiser represents net loss to the radio man, for, unlike the publisher, he gets nothing from the customer direct. A portion of the newspaper's bill is paid by the consumer when he buys his copy. The radio consumer simply flips a switch if the station dares to ask him for help.

And so we have it. The one kind of radio program that suffers very little studio censorship is the all-powerful "commercial." The radio advertiser buys time-space because he wants to sell merchandise. He is not moved by philanthropy, a zeal to educate, a belief in doctrine. He wants to move goods. And it is elementary that people in a good humor are people most likely to be convinced that they should buy something. Even more elementary is the knowledge that repetition makes for familiarity, and that people are inclined to cling to the familiar rather than accept the strange. Hence, from the standpoint of the space-time buyer, the perfect program is one which presents the name of his product in such a way that it will stick in the listener's mind. Of course, if radio programs were one unceasing stream of jokes and cheery little songs, the taste for them would tire. Humor unvaried becomes tiresome like anything else monotonously maintained. Consequently, to secure attention and drill consciousness of his product's name into the skulls of the American public, the radio business psychologist has attempted to stimulate fear, hate, suspense, and concern, always guided so as to arouse the greatest possible emotion without rousing that bugaboo of radio, disgust. For disgust leads to a snapping of the dial to some other point.

The National Association of Broadcasters offers a very impressive code of ethics, of which typical passages direct that no member shall permit the broadcasting of false advertising statements or claims which he knows or believes to be false, deceptive or grossly exaggerated, or defame or disparage a competitor directly or indirectly.[1]

The Columbia Broadcasting System has established an elaborate policy intended to keep the advertiser's interest and the public good as nearly parallel as possible. Especially in the case of children's programs, Columbia has shown a spirit of general progress. It has retained a child psychologist and laid down specifications for treatments of skits. Exaltation of criminals as heroic adventurers, the showing of disrespect for authority, and the presentation of an attractive side to cruelty, greed, smugness, and conceit all are forbidden. Columbia also refuses to accept programs advertising or discussing bodily functions and symptoms of internal disorders, or other matters generally not considered acceptable in social groups.[2] According to the head of Columbia, William S. Paley, it was necessary to reject two million dollars' worth of advertising to put this code in effect, but the public reaction was so beneficial that, Paley said, the loss was made up in short order and Columbia Broadcasting System had to forego an additional two million dollars' worth of new business because all its facilities were exhausted.

The National Broadcasting Company simply makes a generalized declaration against obscenity, which is already illegal, "and all other language of doubtful propriety," and prohibits disparagement of competitors, or the making of false claims, or offensive comments on religions and racial traits.[3]

Undoubtedly, the stations have an earnest desire not to offend the public, for to do so is to repel a listener, but neither do they want to offend time-space buyers, for they are the men with money. But does radio balance fairly the public good and the advertiser's notions of sound entertainment?

George Henry Payne, member of the Federal Communications Commission, in an address to the National Conference on Educational Broadcasting, in Chicago, last November, denounced the type of children's programs then being offered.

The threat to the home through deleterious foods and drugs, indecent programs, nerve-racking children's entertainment, and a sophisticated philosophy that is fundamentally unsound, can only be adequately understood when we realize how long and severe was the struggle to establish the spiritual sanctity of the home. . . .[4]

Commissioner Payne has furnished the authors with copies of letters that came to him from all sections of the country after newspapers printed excerpts from his analysis of radio's faults. They indicate that not all the best advice has been given the sponsors concerning the effect of certain emotions upon the listener.

It is interesting to note that those letters were not from one kind or class alone. Lee De Forest, describing the tone of broadcast material as sinking to a moron level, stated that only the Communications Commission can cause an elevation to proper standards. The evil effects of certain programs were pointed out by neurologists, pediatricians, neuropsychiatrists and general medical practitioners, all asking corrections for the sake of children. A high school history class in New York notified Mr. Payne of its expectation that the commission would force an improvement of educational radio. Parents' leagues, individual mothers and fathers, and casual listeners all entered complaints against programs and commended Mr. Payne's campaign for improvement. If the radio industry gave any serious attention to Commissioner Payne's views, to the response of the people to them, or to other criticisms of radio made at the Chicago conference, none was made evident. The programs continued much in their former vein, and *Variety*, a trade magazine of the amusement industry, reported the advertising agencies and program managers, instead of being chastened, were angry and indignant at the temerity of their critics.

Radio men, said *Variety*, regarded the whole affair as a field holiday for and by swivel-chair secretaries of pressure groups, college professors with a fondness for page one, and other brave-word utterers who consider radio big enough, rich enough, and vulnerable enough to be attacked with a good prospect of newspaper publicity, yet without fear of successful reprisals.

The broadcasters were angry because they felt that the conference was given over to cheap histrionics in some cases, ignorant platitudinizing in others, and marked throughout by lack of fair play and lack of realistic information as to "what radio is and, by law, must be." Just how the law requires or authorizes radio to frighten children and scandalize parents was left to the imagination, but *Variety*, calling the critics seekers after "Genteel Sinecures," declared that station men "see the whole promotion as the brain work of guys who make their living stirring up fussy club women to pass resolutions. It is well within the realm of possibility that radio, as a result of the tomato-throwing at the Drake, will start gathering some ammunition to fire back at the cute kiddies in charge of uplift who make their living by target practice at impersonal enemies that they never expect to retaliate." [5]

Another viewpoint of the broadcasters was held to be that most of the "pressure group executive secretaries and their big-word brethren of the campuses are chiefly vexed that they don't always get choice evening time on the cuff whenever they wish to ballyhoo themselves or their pet projects."

All this outburst was prior to the widely publicized "Mae West incident," in which the National Broadcasting Company was rebuked by the Federal Communications Commission and forced to apologize for a program denounced as indecent, obscene, and sacrilegious. In it the actress performed in a scatological skit concerning Adam and Eve which some fifty stations of the NBC circuit offered without restraint or timidity. The repercussions from this affair were such that many radio people undoubtedly wished they had paid more attention to the Chicago conference and less to Variety which summarized that forewarning of storm as "tripe, bunk," and added that the attempts to improve programs were false conceptions of either authority or powers, or both, on the part of the speakers; that, moreover, in view of the wide publicity given them, the speeches represented libel to the whole industry.

Broadcasters, said Variety, hold that radio is not, never has been, and never will be exclusively for college professors—nor is it for club women. They say that they have made no attempt to please these two classes and do not contemplate doing so in the future, for two reasons. The first is that radio is a commercial proposition, and must therefore appeal to the masses, of which club women and college professors are not even an infinitesimal part.

Their second reason is that licenses call for them to operate for the public interest, convenience, or necessity—which is interpreted to mean for the station men's idea or the public at large, and "not theoretical sideline critics."

Concluded Variety:

Club women came in for particularly bitter condemnation for the way they struck out at kid programs, kid cycles, and kid merchandising. It is the same old criticism, and received the same condemnation; that if the club women would stay home and edit the programs they wanted their children to hear, the alleged offenders would die a natural death.

This is one of the first times in history that the boys have all been so mad together, at the same time, at the same thing. Day after the conference closed, offices and studios were all boiling, with everybody determined to do something, if nothing more than to chase the professors and club women back to their books and their gossiping.

One of the most disastrous results as far as Chicago is concerned is expected to be a prolonged set-back of a heretofore slowly progressing policy for liberality. Both Coast and New York audiences are more tolerant of what may or may not be aired than Chicago—and while the midwest isn't particularly anxious to go in for gags, its writers and producers would like a chance to build situations around triangles and social problems instead of confining themselves exclusively to Cinderella and Gingerbread Man themes.

Some executives are even reticent to cooperate further than necessary with the newly established Chicago Board of Education radio department, not because it was prominent in the flood of criticism, but because it belongs to the same professor classification, psychologically.

It should not be assumed that commercialism in radio is necessarily bad just because of the spirit reported in this article. Indeed, with a proper conception of the public interest, necessity, and convenience, radio financed by advertising can be just as adequate as any other. On occasion it has been.

In times of great emergency, as in the spring floods of 1937 along the valleys of the Mississippi, Ohio, and Monongahela rivers, commercial radio has been the sole source of aid and information to thousands of people. On occasions of state importance, as when a President addresses the nation or a political party holds a national convention, it brings the mass and the masters into intimate relationship.

Radio, built upon advertising and commerce, has acquired its grip upon the American people. It has developed the basis of acceptance upon which it hopes television will extend; and sound radio certainly will have first chance at that extension.

But there is no guarantee. There is no reason why the present holder of a license to broadcast sound should have prior claim upon the improved service if he cannot demonstrate a genuine appreciation of the phrase, "public interest, convenience, or necessity." Again, it is not commercialism but the abuse of license and privilege that has brought criticism upon our current type of broadcasting. And it is the intolerance of the licensees toward critics that provokes the desire to seek withdrawal of licenses from the hands of the abuser who, like the Bourbons, learns nothing and forgets nothing.

12. The Somnolent Cinema

TELEVISION EATS UP LARGE AREAS OF THE SPECTRUM TO THE starvation of other radio services, but that is not the end of its ravening. It threatens to swallow whole industries. Radio set manufacturers will have to transform their technique of production so that they become television set manufacturers. Radio broadcasters must become television broadcasters.

The radio set manufacturers, and the broadcasters who found the commercial band of the spectrum a vein of virgin gold, have recognized full well the danger that confronts them. In regiment formation they have bombarded the Federal Communications Commission to consider their interests as television approaches. They experiment, make treaties among themselves, and offer plans for protection. They might be called sprinters, crouched for the starting gun in a race that will end in fame and fortune for somebody. But among the contestants we see an unwilling fat boy trying to assume the angular position of the ostrich with head in sand. That, in a word, is the way the motion picture industry is behaving as television comes. The bulk of television programs will probably be in the form of motion picture films. For one thing films are more easily televised than stage performances, and have proved so successful that in the present experimental period sixty per cent of the broadcasts are from films. Apart from mechanical perfection there are other considerations. The film story technique lends itself naturally to television; and so does the scenic perfection that the motion picture industry has developed.

But television has a voracious appetite for material. If it comes to operate on a time schedule equal to that of present commercial radio, the present annual production schedule of films will not maintain service for more than three months. To keep up with such a pace the movies will have to undergo radical changes. Present production schedules, if quadrupled, still would not meet the demand. But even if the supply of entertainment can be kept up, the movies may still be reduced to a minor vestigial program service unless a sound bargaining position is established for them. Having undergone one radical change in ownership and financial structure because of unpreparedness, the movie moguls ought by now to be alert to technical change and its threats, but, alas, they seem not to be.

At present the motion picture industry is in two distinct though not entirely separate branches, each dependent upon the other. One branch is concerned with the production and distribution of pictures (Hollywood), the other with exhibition (America). Hollywood concerns itself with studio operation, photography, sound recording, the selection of artists and plots; in a word, with picture creation. Production could go on in a television era, only speeded up or slowed down to meet demand; and nobody outside Hollywood, except those holding stock in movie companies, would know or care.

The exhibitors simply put the finished products before America today and try to ward off the headache which is surely going to overtake them with the advent of television. It would appear as though, when the new consumers are available at the studios, the producers may be in a measure freed from their dependence upon the exhibitor to whom they have had to cater for so many years; but actually the television broadcaster is merely substituted for the exhibitor.

The movie moguls have always been the victims of a mania for, and a complete failure to attain, independence. Before the advent of sound they used their fresh and copious profits to create exhibition outlets of their own wherever possible. Some of these remain today. One of the first ventures into both sides of the market was made by William Fox, a furrier turned nickelodeon operator who acquired a producing company to guarantee his theaters films for exhibition. Fox is a rare character and one of those who make this story possible, for he not only bound production and exhibition together, but overlaid both with sound and with banknotes. At the advent of sound, Fox intensified the chain movement of theaters by pushing the industry into the new technique so that it had to be assured not only of actual distribution of product, but also of equipment in theaters to reproduce programs in a manner becoming to the super-colossal empire that Hollywood conceived itself to be. On the practical side it was recognized that the movies could not go on half silent and half sound. Events and schemes pressed the moguls finally to choose sound.

The arrival of sound movies smashed the structure of such leading companies as Fox, Universal, Paramount, and Radio-Keith-Orpheum, and made them the vassals of bankers. Famous actors and actresses became as obsolete as wooden plows or handmade shoes. Theater orchestras vanished into picket lines; and the legitimate theater became an appendage. Today those few actors who refuse the western adventure find themselves cast in productions which are conceived, designed, and maintained in the sole hope that some film company will take an option on them. Is it inconceivable that the next step in the theater's metamorphosis is a vestigial movie house in which to test public reaction before the great exhibition to the nation by way of the radio spectrum? Will the motion picture theaters occupy the present situation of the legitimate theater? To determine such questions as these the movie industry maintains an institution known as "The Motion Picture Producers and Distributors of America," headed by Will Hays, who was Postmaster General of the United States during the administration of Warren G. Harding.

In 1936 Mr. Hays hired A. Mortimer Prall to make a study of the relation of television to the motion picture industry. Upon learning that this research student was the son of the late Anning Prall (who was then chairman of the Federal Communications Commission, which also had the problem of television under study at that time),[1] one recognizes the astuteness of the "Czar of Hollywood."

Mr. A. Mortimer Prall, in a highly confidential document entitled "Television Survey and Report," advised the movie people that television opens a new and extremely important field for the industry. He pointed out that three times the amount of film they produced would be necessary for television. In addition, "the motion picture industry is composed of great production corporations. They possess every element necessary to the production of the finest programs of sight and sound on film. Writers, composers, artists, designers, architects, engineers, technicians, construction men, studios, special equipment and the world's best actors and actresses are all part of this industry . . . It is clear that the motion picture industry is the only source of supply for television programs."

Two plans were suggested in this report. One was that the present producers apply to the Federal Communications Commission for permission to buy up one of the existing radio chains such as National Broadcasting Company, the Columbia Broadcasting System, or the Mutual Broadcasting System. The other was that the motion picture industry buy up stations not now in one of the four major networks and form a fifth radio chain. That too necessitates application to the commission for license. In other words, he suggested that the motion picture industry engage in the business of radio with the sanction of the commission of which his father was chairman.

There are several obvious faults in this plan. Sound radio is certainly a step towards television. But it must be recalled that television will play in the upper strata of the spectrum. There is, of course, no guarantee by Mr. A. Mortimer Prall that the commission will give the movie industry frequencies for television when the day for commercial exploitation arrives. It could happen that the movie industry would find itself left with two very large and moribund white elephants —the present motion picture studio and theater system, and the sound radio system as well.

Is the exhibitor to be left to his fate by Mr. Prall? This is an important consideration, both for the producers and for the little men with neighborhood theaters. Because of their large investments in exhibition chains it would be suicidal to their capital structure for the great producing systems to allow their theater investments to crash. But however we may pity them we have to ask what incentives there will be for a customer to drive his car, run or even walk to a movie house when his own living room may become a theater; and we can think of none that seems valid. Maybe there are reasons why the movie palace will last despite television. One argument has been advanced to the effect that the theater will remain as a place of assembly because man is naturally gregarious, but that possibility seems a poor comfort to the magnate whose fortune has to depend on it. Rather, he turns to a report of the Academy of Motion Picture Arts and Sciences which differs with Mr. Prall absolutely. It states that all is well and that the motion picture industry has nothing yet to worry about from television.

"There appears no danger that television will burst unexpected on an unprepared motion picture industry," [2] says the Academy, and since that is comfort from his own, the magnate dreams comfortably of *apfelstrudel* and dividends. Whether this is simply whistling in the dark, or is a private word of assurance based on evidence undisclosed to the public, is anybody's guess; but at the risk of destroying peace of mind in Hollywood, we offer as a clue the following clause for a contract that conditions production by ninety per cent of the sound motion picture industry:

No licenses are herein granted or agreed to be granted for any of the following uses or purposes:

THE SOMNOLENT CINEMA 127

(1) For any uses in or in connection with a telephone, telegraph or radio system or in connection with any apparatus operating by radio-frequency * or carrier currents. . . .³

Television can operate only on radio frequencies, or on carrier currents through wire cables. This clause is a part of the contracts between the American Telephone and Telegraph Company and seven of the eight major producers of pictures in Hollywood. Have the movie men been assured by their masters that television will be allowed to develop only as the masters will? Or have they overlooked that clause entirely and simply concluded that movies have their place in the world and can't be shaken out of it? We cannot but succumb to our habit of quoting official documents as a means of showing that there is more than guesswork and intuition behind the warning that the movies may be on their way to extinction or absorption. Bear with us in a flashback of history concerning the sad story of the silent film and the sound machine. It is told briefly in two excerpts from the memoranda of a memorable character whom we shall identify shortly. He, more than any other, drove the nails in the coffin for Gene Fowler's fabulous "Father Goose." Here is memorandum number one:

The motion picture industry in the United States owes us about sixteen million dollars and our expected revenues from the industry for the next ten years is about sixty-five million dollars. This is a large stake and establishes our interest in the welfare of the motion picture industry.

* In the first sound recording contracts between the Bell telephone system and the Vitaphone Corporation, television was specifically mentioned, but in characteristic fashion this was withdrawn as events and legal stipulations came near toward conflict.

The industry is in a serious financial condition and some of the large companies are faced with possible receiverships. The morale of the management in many instances has been greatly lowered. Unwise remedies are being applied and reorganization efforts are being made that in all probability will not be successful. As a result of these conditions our stake is in jeopardy.

We are the second largest financial interest in the motion picture industry. Our stake is next to that of the Chase Bank....

I believe that the protection of our interests in the motion picture industry requires that we should have authoritative conferences with the Chase Bank at the present time. Our interest should be made clear and our influence felt. We can do things the Chase cannot do in the interest of the common good and Chase can do things we cannot do....[4]

Number One was written on November 5, 1932.
Number Two:

I have also had innumerable proposals that ERPI go into this or that phase of the motion picture business. These I have declined without bringing to your attention because I recognize such proposals to be contrary to the Bell system policies and interests, and even though they offered ERPI opportunities for advantage and benefit. It is true today, as it has been for three or four years, that the Telephone Company can control the motion picture industry through ERPI without investing any more money than it now has invested.

I am not recommending that this be done, even though I know that the salvation of the picture industry lies in this direction. The industry is in crying need of the kind of strength and character that could be obtained through the influence of the Telephone Company.[5]

Number Two was written December 7, 1933.

Had "this direction," as described in the correspondence between J. E. Otterson and E. S. Bloom, officials of the American Telephone and Telegraph Company, been followed, all of the motion picture industry would soon have found itself under a single management, with a single studio operating organization and turning out pictures to be sold and exhibited through apparently competing sales systems. And, according to most standards of artistry and theatrical enterprise, disastrous effects upon the movies as entertainment would have been invited thereby.

It is crystal clear that only the judgment of its distant financial masters left the motion picture industry a figment of independence when it tottered under the impact of sound technique. That figment of independence has been nourished carefully since, but never enough to allow the original moguls to re-establish themselves completely.

Let us remember and never forget that of the eight major producing companies, seven are bound up so that they cannot sell or lease their films for television if they want to; and that is why, perhaps, the Academy of Motion Picture Arts and Sciences recommends no fears. They put their faith in the cool judgment of the financiers far away to ward off the new threat. But what of the eighth major producer? And what of that great industrial magic, Competition?

The telephone system moved in on the motion picture industry with a new technology, the sound films, and tied up ninety per cent of production with its contracts. Of the remaining ten per cent, the apparent competitive fringe, virtually all fell into the hands of the Radio Corporation of

America, which proposes itself to be the perennial nemesis of the wired communications services.

And not too unsuccessfully, as witness this further memorandum by an A. T. & T. Company official:

> In the talking motion picture field they [RCA] are competing very actively with us at present, as you know, to develop an affiliation with the large motion picture producers, and competition between us all will doubtless ultimately result in a situation highly favorable to the motion picture interests and opposed to our own.
>
> This is an extensive and highly profitable field and it is quite worth our while to go a long way toward making it practically an exclusive field. I believe that we could justify from a commercial standpoint paying a large price for the liquidation of the Radio Corporation for this purpose alone.[6]

The author of this remarkable view was by no means foolish. Events show that he saw correctly the problems of protecting vested interests in times of technological change. And perhaps it is because the motion picture producers realize that they are really in no position of command just now that they cower like white rabbits as events start their march again. But what about the movies' masters?

13. No. 195 Broadway

THERE IS ONE ORGANIZATION WHICH HAS MISSED NO OPPORtunity to prepare for television. Its influences permeate finance, engineering, sociology, law and that peculiar field of operations known as "public relations." Its sentinels and intelligence operatives are both able and alert, and they are very, very many.

The institution in question is the American Telephone and Telegraph Company, of No. 195 Broadway, New York City, the biggest business corporation in the world. It is known to the general public as "the A. T. & T. Company" and to some inflamed politicians as "the octopus," but the officials who make it go have a much more apt and glowing phrase, "the Bell system." This Bell system is devoted wholeheartedly, but not exclusively, to the development of domestic telephony within the United States.

Its economists know, for instance, that the future of television is their own future, and have long since made plans intended to insure that nothing bad shall happen to the Bell system. On that account we are entitled to examine the telephonic octopus, note its habits and philosophy, and make deductions as to its probable behavior in the present issue.

The Bell system today has assets valued in excess of five

billion dollars.[1] The principal structure is composed of the A. T. & T. Company, twenty-four "associated companies" operating domestic telephony in the United States either as wholly owned or controlled licensees, and four associated companies operating transoceanic radio telephony. Between them they link together 92.77 per cent of all the telephones in the world.[2] Within the United States proper, the Bell system has a direct property equity and profit interest in almost exactly the same percentage of the total American telephone system. Of the more than eighteen million telephones in service in the country, a bare seventeen thousand fail to connect with its network.[3]

Is telephony the only portion of the Bell system which may be termed "gigantic"? Here are some further aspects:

The Bell system dominates sound broadcasting, in that its network of wires is the only one over which national chain programs can travel adequately. It dominates both production and exhibition of sound motion pictures, international radio telephony, wire transmission of news and news pictures, teletypewriting and teletypesetting, and terminal apparatus for submarine cable systems.

It either controls or maintains a potent interest in the making of electro-surgical knives, medical diathermic devices, watch testing and race timing apparatus, instruments for the hard of hearing, radio transmitters, public address systems, phonograph instruments, police radio equipment, turntables for funeral parlors, and a multitude of other devices, methods, practices, and operations encompassed by the nine thousand patents it holds from the United States Government. Its influence and interest extend even further into companies with which it has reached an agreement on competitive activity allowing for specific licensed use of six thousand additional patents.[4]

One of the characteristics of the Bell system most often spoken of with pride is that no single stockholder owns as much as one per cent of the total issue. In sum its stockholders number approximately 715,000 who are happy, year after year, to accept their dividends of $9.00 per share, without thought of a meeting at which to hear reports from management or consider voting some officer out of control. Where, indeed, could the 715,000 meet if they were so minded? Not in the Yankee Stadium at New York. It holds only 80,000. Not in Soldiers' Field at Chicago. It holds only 105,000. Not in any communal meeting place now conceivable. No American city in the year 1937 was designed even to house and feed 715,000 visitors en bloc, let alone accommodate them with a place for hearing reports and acting on motions. But regardless of these absentee owners who are divested literally of all management power, the Bell system, most people admit, does as good a job of domestic telephony as we can imagine, especially since there is no competitor offering to raise the standard.

Indeed, the Bell telephone service is of such caliber that it appears a plan was once instituted to have the A. T. & T. Company take over the system of a foreign nation in which telephony, as a part of the national defense, is a governmental monopoly considered of the greatest strategic importance. So unusual was this proposal, considering the rank and power of the country in question and the influence of the negotiator for the change, that we quote in full the following cablegram, dated August 1, 1928, from a Bell system representative:

MR. E. S. BLOOM

OUR TELEGRAM 732 LORD BEAVERBROOK INFORMS ME HE THINKS TIME IS RIPE FOR THE AMERICAN TELEPHONE AND TELEGRAPH COMPANY TO TAKE OVER THE ENGLISH TELEPHONE SYSTEM AND THAT THERE IS GENERAL DISSATISFACTION WITH GOVERNMENT ADMINISTRATION WHICH HAS TAKEN POLITICAL FORM AND WHICH HAS BEEN AGITATED IN NEWSPAPERS AND POLITICAL CIRCLES STOP BEAVERBROOKS PAPERS ADVOCATE PRIVATE OWNERSHIP AND FAVOUR AMERICAN TELEPHONE AND TELEGRAPH COMPANY STOP HE WOULD NOT BE SURPRISED IF THE AMERICAN TELEPHONE AND TELEGRAPH COMPANY WERE INVITED AT EARLY DATE TO MAKE SURVEY FOR TAKING OVER SYSTEM STOP WILL HAVE FURTHER DISCUSSION WITH HIM AND ADVISE YOU IF ANYTHING DEVELOPS

OTTERSON [5]

Upon what authority from the British Government Lord Beaverbrook may have been operating, no record shows. Nor is it shown whether the Bell system made any effort to extend its system into the British Empire. All we know is that the Postmaster General still operates telephony and radio in England.

Perhaps the Bell system managers felt they have enough to do at home. Radio, we know from the hints thrown out by Mr. Sarnoff of the RCA, is willing to give the wire companies competition at any time. He recently stated his eagerness for the encounter:

The ideal way of sending messages is to hold up a printed sheet that will be immediately reproduced at the other end; facsimile transmission and television are about ready [for that].

If a strong, unified telegraph company was to be in the field, the telephone people would be in about the same situation the telegraph groups now find themselves in.[6]

Mr. Sarnoff's organization is the Bell system's chief competitor for first place honors in television. How does the management of the Bell system propose to protect its 715,000 absentee owners against loss of $9.00 per annum per share, and itself against the sort of threat to which Mr. Sarnoff gives words? The gentlemen who manage the Bell system are well aware that they have a tradition of capability and rationality to maintain. That tradition is as old as the Bell system itself. It was wrought in a hard school, and it has endured because Bell management has never lost the knack of fusing luck with action, knowledge with aptitude, and of keeping a steady focus upon the main chance. How completely aware management is of the perpetual war between devices of communication may be gathered from the following memorandum:

A primary purpose of the A. T. & T. Company is the defense and maintenance of its position in the telephone field of the United States. Undertakings and policies must be made to conform to the accomplishment of this purpose.

The A. T. & T. Company is surrounded by potentially competitive interests which may in some manner or degree intrude upon the telephone field. The problem is to prevent this intrusion.

These interests are characterized by the General Electric Company, representing the power and light group, the Radio Corporation of America, representing the radio group, the Western Union Telegraph Company, representing the telegraph group, and the International Telephone and Telegraph Company, representing foreign telephone interests. Other miscellaneous interests which may not fall in any one of these groups may appear as potential competitors at any time but the consideration can be confined to these four groups as illustrative of the whole. . . .

Each of these large interests is engaged in development and research that is productive of results that have an application outside of their direct and exclusive field.

On the whole, it seems to be essential to the accomplishment of the A. T. & T. Company's primary purpose of the defensive protection of its dominating position in the domestic telephone field that it shall maintain an active offensive in the "no man's land" lying between it and potentially competitive interests.[7]

These are excerpts from a document famous within the inner circles of high finance and telephony. It is known as the "four square memo" prepared on January 13, 1927, by J. E. Otterson, author of the Beaverbrook cablegram, writer of the memoranda quoted in the preceding chapter concerning how to save the movie industry, and executive vice-president of Electric Research Products, Inc., a corporation set up fourteen days before to exploit the by-products of the Bell system's laboratories in general, and, specifically, to make profitable its inventions in the sound motion picture field. Five months after he wrote the "four square memo," Otterson was made president of ERPI, as the exploitation company has become known.

In Otterson's memoranda and his actions (he was said by *Time* to run ERPI with "battleship efficiency") the Bell system's powerfully effective trading philosophy is expressed, though lacking in two vital qualities. Otterson had not the knack of negotiation and personable dissimulation.

As a midshipman at the United States Naval Academy, he was trained to execute orders exactly, to state objectives literally, and to keep his mind wholly on ultimate effects. He was never, in the Bell system, able to divest himself of these worthy but narrow qualities. As a friendly observer of his actions expressed it, one could order Otterson to capture Manila and feel confident that he would do so. But, as the record shows, one had no reason for comfortable assurance that he would not, in the taking, stir up all the world's neutrals and rouse a coalition for retribution and punitive counter-attack. In fact his handling of the ERPI was exactly that kind of a campaign. It cost him and his employers dear.

But add to Otterson's "four square memo" the practical doctrine of avoiding open encounters and offering settlements in lieu of suits, the gentle art of using soft words rather than harsh threats, and one has in whole that fascinating concept known as the Bell system's trading philosophy, of which there was need on Monday, February 14, 1876, when a bewhiskered American known as Elisha Gray rushed into the department of electricity in the U. S. Patent Office to file a caveat, warning the world that he was about to produce a device he called the telephone. He promised, thereby, to transmit the sound of the human voice from one point to another. Mr. Gray's entry on the records of that day's business was No. 39. He was shocked to learn, next morning, that entry No. 5 was an application for patent by one Alexander Graham Bell, a native of Scotland who had become an inventor after arriving in the United States. Mr. Bell had promised to deliver by wire not the spoken voice, but only sounds and noises. That was enough.[8]

Bitter years of dispute and legal controversy followed. Reputations were attacked, fortunes spent. Had the clerk who recorded the entries of February 14, 1876, done so on the basis of the hour they were received, there might today be no "Bell system," however apropos the title to the clangor of the instrument. Instead, there would perhaps have been a "Gray system." One could spend a lifetime musing upon the fanciful enticements to reason offered by counsel who wrangled for priority in the telephone patent position. One of the more ingenious ran to the effect that the patent clerk stacked the applications on top of each other as they came, so that the last to be filed was the first to be recorded. Ergo, Bell's application, being No. 5 on the listing, was fifth from the last to be received and Gray's, No. 39, was filed much earlier in the day.

But we are not now interested in tortuous legalism, nor even in the famous "Watson, come here; I want you," which, incidentally, was achieved after the Bell patent was filed. The important thing is that out of the welter of suits, caveats, claims, and acrimony, Alexander Graham Bell and associates emerged in control of telephony; and we observe that they emerged by way of the Bell trading philosophy.

14. The Bell System

THE FIRST STEP TOWARD DEVELOPMENT OF CONTROL OVER the new art was the formation of the Bell Patent Association on February 27, 1875, in Boston, Massachusetts, by Alexander Graham Bell, Thomas Sanders and Gardiner G. Hubbard.[1] It is at least as much to Hubbard as to Bell that the system owes its dominance today. Sanders and Hubbard agreed to underwrite Bell's experimentation, in return for which the three would split equally the stock ownership for control and management of patents.

On September 1, 1876, the three Association members hired Bell's assistant, Thomas A. Watson, to work on experiments and make telephones, at a salary of $3.00 a day, and a promise of one tenth interest in all patents upon developments by a joint stock company.[2] Any inventions perfected by Watson, Hubbard decreed, would belong to the Association, a prime principle of operation in effect today throughout the Bell system.

Shortly thereafter, having filed four basic patent applications, the Bell system of 1876 made a gesture of surrender to a powerful competitor. It offered to settle everything for cash and go out of business. At that moment, the Bell system was weak and undeveloped. It was deeply involved in patent suits with Gray, and was further threatened by the

Western Union Telegraph Company, which had belatedly taken an interest in the telephone. But luck was with the Bostonians.

Western Union, emerging from the Civil War as the first great network of electronic communication, viewed the telephone as a toy when the Gray-Bell patent litigation began. Furthermore, its own management was harassed by what appeared to be more important affairs than progress with the wholly experimental new device from Boston. William H. Vanderbilt was chief owner and manager of Western Union, and Jay Gould was battling him up and down the length of Wall Street for supremacy. Gould won, eventually, but the effect of the fight was Western Union's undoing, and meant the driving of telegraphy into obsolescence by telephony, for his attention was diverted from the new invention in the vital formative period from 1874 to 1876.

Not until 1877 did Western Union decide to take a serious interest in telephony. And when it did, with its fatal proclivity for backing the wrong horse it chose to buy up Gray's patents and not those of the Bell Patent Association, which had been proffered for $100,000 by Hubbard in his moment of hesitation a few months before.[3] To Gray's, Western Union added the patents of E. A. Dolbear, another early telephone inventor, and the confections of its own almost exclusive genius, Thomas A. Edison. Why should it bother, then, with the Bell business from Boston? A fatal error, and one that we find the telephone monopoly has never made.

When one reflects that from the fall of 1876 the star of the Bell system has been resplendently rising and the star of

Western Union has followed an erratic downward course, the trading philosophy expounded by Gardiner G. Hubbard and glorified by the management oligarchy thereafter acquires even richer significance. For, seeing they were in for stormy weather, the little Bell contingent capitalized, on July 9, 1877, the Bell Telephone Association, with seven stockholders and with Gardiner G. Hubbard as trustee, and set out to do business.[4] Thereupon Mr. Hubbard laid down the second great Bell principle: lease instruments and license their use, but never sell.

To that principle, as to the one involving patent rights, the great Bell system adheres today wherever possible; and there is little chance that it will follow any other in the case of television if allowed its own way. The telephone subscriber pays rent for his telephone but does not own it; the motion picture producer is in the same predicament. So is the exhibitor. So is the newspaper proprietor buying telephoto pictures. So is almost every user of Bell services, devices, or equipment except in the very rare instances of organizations prohibited by the terms of their endowment from such leasing.

In the latter part of 1878, as Western Union began to take a serious interest in the "toy" and began to push its own Gold and Stock Company in the manufacture and distribution of telephones, the Bell system reached out in the conventional way for new capital. It then developed another characteristic that continues to this day: association with the best names in finance and with the public conception of all that is good and noble ("sound public relations"). Herein Sanders was of value. He, as treasurer of the company, brought around investors with names then as now potent in Bostonian and American financial and social life—Bradleys, Saltonstalls, a Forbes, a Carlton, a Fay, and a Silsbee. They organized the New England Telephone Company to pour capital into development of the Bell system, and in return received under Hubbard's terms an operating license under the Bell patents.

Out of the licensing of the New England Telephone Company evolved still another great Bell principle later expressed succinctly by Theodore N. Vail, President of the A. T. & T. Company, and himself a shining example of Hubbard's astuteness. Vail came to the Bell system in 1877, surrendering a good government job as superintendent of the railway mail service, for, as a friend put it, "a damned Yankee notion, a piece of wire, with two Texas steer horns attached to the ends, with an arrangement to make the concern bleat like a calf." [5]

In 1878 Emil Berliner, a Jewish dry goods clerk in Washington, D. C., with no special training in engineering, patented a telephone transmitter which to this day is the basic structure of the microphone. When the Berliner caveat was filed, Vail did not foresee necessarily its implied powers, but he knew better than to reject a possibility. So did the Bell system when it was offered the De Forest audion; and today, as thousands of amateurs scrawl away on pads with pencil in attics, dark corners, and cellars, and tinker with improbable radio hookups, it is still alert to their potential threats.

Vail was an apt pupil at the feet of Hubbard, and came in time to see not only the wisdom of that Yankee trader in doing business, but also to state the true principle of effective communication by telephone or any other electronic system:

> You see, in the first place our idea of the development of the business was that there was a system to be developed that had in itself a value far beyond anything that might be called a mere patent business. Our idea was to get control of that as a permanent thing. . . .[6]

It was to "get control of that as a permanent thing"—the business, whether it should extend to movies, diathermy, or television—that the Bell system set about, under the direction of Hubbard and Vail, to consolidate and link irrevocably the independent systems which began to sprout as invention made the telephone seem a necessity.

On March 13, 1879, the steady progress in corporate development took a new turn with the formation of the National Bell Telephone Company.[7] (Already, within three years, they were speaking of "national.") The National Bell assumed all the benefits and powers of prior organizations and set out to clean up the field.

First, it settled with Western Union. On November 10, 1879, an historic treaty in the communications war was signed.[8] Like most treaties it was not, in its effect, mutually preservative of the signatories. In that agreement, the Western Union gave over to National Bell all the apparatus, patents, and rights of the Western Union telephone subsidiary, American Speaking Telephone Company. It further agreed to stay out of the telephonic field, presumably forever. In return, National Bell agreed to pay Western Union twenty per cent of its income from telephone rentals and to stay forever out of telegraphy. In later years, when the A. T.

& T. Company came to operate a gigantic leased wire system for teletypewriter transmission of news and other printed data, Western Union sought, according to report, to invoke the treaty of 1879. It was about as effective as the Kellogg-Briand Pact has been in halting Japanese activity in Asia.

That assignment of twenty per cent of royalties was a heavy sacrifice for the Bell system, but it demonstrated the inherent genius that has characterized the Bell line of unbroken success, for it established peace with a major competitor and allowed Vail time to extend his licenses, knit together long distance service, and make the Bell system valuable not solely for patent rights alone. It preserved the growing organization against the day in 1894 which he knew must come; the day when the patents would expire and the original Bell method would, under the Constitution, be declared a part of the public domain.

On April 17, 1880, in accordance with a special act of the Massachusetts Legislature, the American Bell Telephone Company was formed to knit independents inextricably into the Bell system.[9] It is historically of importance, in seeking the outlook for television, to note the actual devices by means of which the Bell system met the independent competitors, once it had conciliated and bought off its one most powerful obstructor.

On October 20, 1880, the Bell Company brought suit against the Peoples Company, of New York, and carried that basic case to the United States Supreme Court, which, on March 19, 1888, rendered a four to three decision in favor of the Bell system. In all, during the seventeen years of its original patents' existence, the Bell system brought more than six hundred infringement suits. Only a few ever reached court. Most of the independents surrendered either to cash, the offer of excellent service, or in the face of unpleasant alternatives. But soon the gentlemanly genius was forced to extend itself, to drop the manner of sue-and-settle. Then, as once later, the Bell system became openly belligerent, and its existence was endangered by aroused public and other interests.

Television enthusiasts seeking to read the future will be interested to learn that after the Bell system lost patent control, the Bell officers defended their position by active propaganda campaigns against the independents: by refusal to connect for service, and refusal to sell Bell instruments to non-Bell companies; by attempts to eliminate outside help for independents, either financial or technical; by cross-licensing and exclusive purchases of "key" units within attempted independent combines; by loans to stockholders in independents, with stock in the independent taken as collateral; and, of course, by purchases behind dummy masks of controlling interests in independents.[10]

Theodore Vail, having retired in 1887, was not with the telephone company in that period of open combat, when, as the propaganda and financial buccaneering campaigns reached their peak intensity under the administration of President F. P. Fish, the Bell system appeared headed for serious trouble. In 1894 there was strong pamphleteering for government ownership. Under Fish's administration ill-will toward the Bell system grew bitter as it has been bitter only once since—during the battles of J. E. Otterson to gain control of the sound motion picture industry for the Bell system's Electrical Research Products Company, Inc. Fish put the A. T. & T. in a dominant position, though in so doing he threatened its whole existence. Otterson similarly imperiled it, even as he proudly wrote, in his report in 1933, that "it is true today, as it has been for three or four years, that the Telephone company can control the motion picture industry through ERPI without investing any more money than it has now invested."

Under Fish's management it was determined that the Bell system's ownership should be sown, like oats before an October gale, across the fields of America. All of Wall Street was eager as Fish made ready. But he was too unpopular, had prejudiced too many minds. The Bell system must preserve the manner of the gentleman. And so Fish was retired. The corporate blanket was shaken in 1907 and out rolled a new American Telephone and Telegraph Company under the presidency of Theodore N. Vail, who had been persuaded to return to control and give management a good name.[11]

The position of the telephone company just prior to that action was, to put it concisely, that it had to choose whether to remain small and independent or grow large and surrender to banker control. It chose the latter course, and the bankers were the ones who dictated Vail's return to command. As a further stipulation, George F. Baker and J. I. Waterbury were put on the board of directors.[12] The sale of the A. T. & T. stock was carried out with great success by what was known as the Morgan-Baker group in Wall Street. Vail was identified as a "Morgan man." His successor as president of the A. T. & T., H. B. Thayer, bore the same subtitle. To this day, in financial circles, the Bell system is

THE BELL SYSTEM 147

considered a "Morgan outfit," just as the Radio Corporation of America is called a "Rockefeller outfit."

It is difficult to say how far the Morgan interest in the telephone company really extends; but there can be no escape from the conclusion that relations between the Bell management and the Morgan Bank are more than cordial. In the case of the Rockefeller connection with RCA, the facts do not appear of record, and it is probable that there is more guesswork than knowledge back of the common Wall street gossip that "Rockefeller's running RCA, now."

There is a curious fatefulness about turning points in the Bell system's history. For on Christmas Eve, 1906, even as the bankers were preparing for the new stock issue, operators of Marconi's still not entirely accustomed wireless telegraphy device were startled to hear, in earphones built to give off the dots and dashes of the Morse code, a woman singing, and then a man asking that all who had heard the first wireless voice telephonic broadcast please write R. A. Fessenden, at Brant Rock, Massachusetts.[13] Thus are sown the seeds of obsolescence. And it is because those seeds are sown so easily, so innocently, so wholly without warning, that the Bell system remains ever alert, operating in a no-man's land, restlessly at war with all the world, seeking universality of acceptance behind a mask of politeness, gentlemanly appreciation, and honest intent to operate within the framework of the law.

Management cannot forget the casual advent of radio telephony; nor can it forget that the American Telephone and Telegraph Company bought thirty per cent of the stock in Western Union in 1909, a controlling interest that elected Theodore N. Vail president of Western Union while president of the Bell system's central corporation. That dominance [14] over Western Union was broken only by an anti-trust action of the Federal Government in 1913. Neither can it forget that Bell system money financed sound motion pictures.

The Bell system today invents, manufactures, leases. It is a system in which ownership is wholly divested of control, in which management is wholly an oligarchy determined to perpetuate itself by supremely successful administration. It is a system which, through the power of its finances, the multiplicity of its patents and licensing agreements, through the prestige and ramified connections of its directors, would seem to be described most justly in the cautious words of the Federal Communications Commission:

> American Telephone and Telegraph Company and its many subsidiaries and affiliated companies command a strategic position in the social, educational, economic and political life of the American people.
> To an unprecedented degree in the world's history, communication of intelligence by word or picture, and to some extent printed news, is under the control and surveyance of a single private interest.[15]

The Bell system, in a word, is a very clever colossus, alertly poised to make the most of every golden moment.

15. The Belle of Hollywood

JUST WHAT MAY BE EXPECTED OF THE BELL SYSTEM BY THOSE who want to wrestle with it for television dominance can be guessed by a quick review of what has happened to competition in two highly profitable adventures into nontelephonic fields. Twenty years after Fessenden's historic broadcast from Brant Rock, Massachusetts, radio was clearly an industry of competitive importance. So, also, was the motion picture industry. Neither could be described as having any direct connection with the "Yankee notion" that had fascinated Theodore N. Vail and the rest of the world. Yet together they represent the major components of television.

The Bell system, in 1926, had no competitors in telephony. It was dominant and solid. But the philosophy of stifling competition before it could come into being, of buying up and weeding out,—the trading philosophy so ably summarized in Otterson's "four square memo,"—would not let it rest. The story of the Bell system and radio will be told in detail further along, with the emphasis on Bell's principal competitor, the Radio Corporation of America; but the way the telephone industry brought out sound movies must be told now. It is a Bell thing, pure and simple.

Inventors invent and nobody quite knows how to stop them or why they should be stopped. The Bell theory has always been to use inventions in nontelephonic fields, first, to ward off potential competition, and second, to bring in whatever profit is available . . . but first and last, to ward off competition.

In 1926, the Bell system had on its hands a system of phonographic disk recording which it considered feasible for synchronizing with motion pictures. At that time the Bell Laboratories were a year old and Electric Research Products, Incorporated, was not formed. Western Electric was the great manufacturing subsidiary of A. T. & T., and Western attempted to market all scientific discoveries directly, but without great success. A contractual letter was given to one Walter J. Rich, a promoter, authorizing him to license motion picture companies to use the Western Electric sound recording devices, and exhibitors to use the Western Electric sound machines in their theaters.[1] Rich and the seven busy Warner Brothers formed a corporation they called Vitaphone, wholly owned between themselves, and the upshot was that Western Electric made Vitaphone its sole licensee, with power to grant sublicenses to other movie producers and the exhibitors.

Out came several short subjects—Raquel Meller singing, Eddie Peabody playing his banjo and leaping wildly, and that full length triumph, Al Jolson in *The Jazz Singer*. A new art was in the making . . . and an industry in terror. Inventors popped out from everywhere with claims of better, prior devices. Nobody will ever know whether they were right for they were lost as the Bell system moved steadily into power. The movie producers, scared as they were, refused to sign up with Warner Brothers. Only one sublicensee was acquired, William J. Fox, the onetime furrier who could see a good thing from afar. The Fox-Case Corporation got in on the ground floor, and it stayed there, as subsequent events will show.

On December 30, 1926, Western Electric formed Electrical Research Products, Incorporated, and handed over its operations to Otterson, with just what instructions the public has never known. Otterson, ever the direct action man, strained relations with the Warner Brothers and Rich, and brought out a new agreement in which Vitaphone was relegated to the position of a mere licensee along with the rest of the boys—if the boys would only sign. Otterson found William J. Fox more congenial than the Warners, and more imaginative—or perhaps we should say more apparently congenial, and a great deal more imaginative.

The other movie producers had imaginations, too. They could see themselves involved with patent suits, poor equipment; could see obsolescence advancing upon great investments at rapid speed, and everywhere upon the wall, in large letters, the shocking word "bankruptcy."

In February, 1927, the majority of them huddled together and brought forth what was known as "the Big Five non-action agreement," according to which none would install any kind of sound movie systems for one year, pending a decision as to what single type should be adopted throughout the industry as most free of patent liability, most perfect technically, and most likely to succeed financially.[2]

Just what went on between the Bell system's Mr. Otterson and the movie producers of America during the year 1927 has never been told. William J. Fox has disclosed somewhat of his own connections with Otterson, in a best-seller entitled *Upton Sinclair Presents William Fox;* but not until Will Hays, the "Czar of Hollywood," and Louis B. Mayer, the "Ambassador of Filmland" and Sydney Kent and others take up the confessional pen will it ever become known just why and how, at the end of the "non-action agreement," four of the five signatories to the "Big Five" pact agreed that Western Electric was their choice.

One recalcitrant, no doubt feeling somewhat as Japan did when the League of Nations was presented with the Lytton Report on the situation in Manchuria after the "Shanghai Incident of 1931–32," chose to withdraw. That one was Radio-Keith-Orpheum, and it voted to go its own way with the Bell system's growing competitor, the Radio Corporation of America, which had brought out a system known as "Photophone," a sound-on-film. The fact that RCA was parent to RKO may have had something to do with that decision.

But Mr. Otterson and his close companion, Mr. Fox, were happy enough that ninety per cent of the motion picture industry realized its better interests and signed itself over to the Bell system on contracts to use Western Electric equipment for a term extending to 1944. These producers who owned exhibiting theaters signed at the same time to use Western Electric reproduction systems also. The terms of those contracts were: [3] that all equipment was furnished on a lease basis only, and that no competitive equipment adjudged inferior to the Western Electric type could be used for reproduction of film made with a Bell sound system.

Royalties were fixed at not less than $100,000 a year, chargeable at the rate of $100 per thousand lineal feet of negative film cut for release printing, in the case of newsreels, and $500 per thousand lineal feet in the case of feature production or other studio products. Producers were required to pay all those royalties, regardless of whether they might use Bell systems or those of competitors, if the competitor's equipment or technique should embody any system or instrument covered by a Bell patent. Finally, *the equipment was not to be used in conjunction with a wireless frequency, carrier current, radio apparatus—or in television.*

An immediate uproar centered around clauses two and three, of course. Number four has been until now a "sleeper" waiting the day of surprise and pain. When the great nonaction pact resolved into surrender to the Bell system, producers and exhibitors were bound to meet ERPI tests on whether any competitive devices were to be operated in lieu of, or in conjunction with, the Bell systems in movie companies with ERPI sound licenses.

The yardstick of "equal quality and volume" was whose yardstick? Nowhere was it stated what the standard of value would be. A growing number of small, would-be competitors in the making of sound devices yelled for ERPI to state its tests. RCA and its Photophone organization were particularly vehement. But Mr. Otterson, the direct and straightforward fighting man, became vague, gazed ceilingward, and consulted his engineers and lawyers interminably. First it was a technical problem. Then it was a legal problem. In the one office he met with competitors who demanded that ERPI state what it considered "equal quality and volume" tests in recording and reproducing instruments. In the other office, he met with Mr. William J. Fox.

Finally, Paul D. Cravath, counsel for RCA, went direct

to Walter S. Gifford, president of A. T. & T. Otterson's tactics of obfuscation, he said, appeared to RCA and Cravath, deGersdorff, Swaine & Wood as illegal; but before instituting suit he just wanted to know whether "our information is correct." [4] The RCA suit against ERPI was never filed. What passed between Mr. Cravath and Mr. Gifford remained, like Mr. Otterson's conversation of 1927 with Louis B. Mayer and Will Hays, "off the record." But ERPI dropped its strictures upon interchanging Bell and competitive equipment.

In the matter of royalties, ERPI's famous contracts worked out so that any Bell licensee, whether he used Bell equipment or some other qualified as "just as good" from a patent standpoint, would be forced to pay the usual Bell royalties. In other words, being licensed by Bell, he must pay the standard Bell royalty on every foot of film, even if he should use a system offered by RCA. This obtained, despite the fact that under a cross licensing agreement of 1926 both parties (RCA and A. T. & T.) had rights, royalty free, to use each other's patents.

Again ERPI stood its ground until David Sarnoff threatened war in the courts.[5] Slight concessions were made, but the row dragged along until 1935, with RCA actually going so far as to send over to ERPI, "for examination by your counsel," the draft of a suit charging ERPI with violation of the Sherman Anti-Trust Act, the Clayton Act, and demanding triple damages.[6] On December 19, 1935, ERPI abdicated on double royalty provisions and simply stood on the basis of technical perfection and the status quo of their contracts to keep a superior position. What happened, in effect, was that a monopoly was replaced by a duopoly, for others were still restricted by the "double royalty" provision. The clause was by no means removed: it was only declared inoperative against RCA.

Just why the Federal Government never made vigorous effort to act upon the RCA charges of anti-trust law violations by ERPI has never been disclosed. The Federal Communications Commission, from 1935 to 1938, spent $1,500,000 in a special investigation of the Bell system, and obtained copies of RCA's proposed anti-trust action, but so far the Department of Justice has apparently failed to investigate or make any recommendations concerning them. If they were baseless arguments why did RCA bring them? Why did ERPI surrender at RCA's demand its valuable double-royalty principle of collection from licensees using other equipment than its own? If the suit had any merit, why did the Government neglect to investigate?

At any rate, the Bell system is now inextricably woven into the production of motion pictures and their exhibition, and well placed for influence upon television because of the terms of the producers' license prohibiting use of Bell sound-made film for television exhibition.

And so we go back to the matter of Mr. Otterson and Mr. Fox. Naturally, as sound equipment was installed throughout America's movie theaters, small and large manufacturers fought for the chance to supply equipment and do repair work.

ERPI first fought openly and made picture exhibitors accept compulsory ERPI service and repairs, replacement and techniques, even though it was admitted within ERPI that undesired "service" was being forced upon the "customers." Finally, the victims began to organize for collective defense, and so the compulsory service system eventually was waived; but exhibitors who might venture to use a competitor's products or repair systems were required, upon returning to "Old Reliable" for good service, to pay $35 a day per man, plus expenses, for emergency repair of a Bell owned sound system put out of order while any competitive part or repair system was in use.[7] This special fee was greatly in excess of the regular ERPI service charge, and it had its effect upon exhibitors, who, like everybody else, were interested in fights only so long as there was a chance of profitable victory. It ought to be added that ERPI had a good argument for the compulsory service system. Theatrical exhibitions were continuous. A breakdown was not only bad for the exhibitor but for the Bell system, which might be blamed. Television set owners may draw a moral from that. The unanswered question was whether a Bell system of sound reproduction would break down more often alone or in combination with competitors' devices than would a competitor's, standing by itself. The record shows no test of comparative ability by an external adjudicating body. The Federal Communications Commission can perform that function in the case of television, however.

Curiously enough, the "little fellows" of motion pictures showed more fight against ERPI than did RCA. Inventors, manufacturers of competitive equipment and parts, and licensees banded together and brought a series of more than twenty suits against ERPI. Most of these have been settled out of court, with neither party claiming a victory. But by then ERPI had served its primary function for the Bell system, and Otterson had been advanced, in June of 1935, to supreme power in Paramount Pictures—only to be thrown out in a fast, revengeful deal by the real movie men, who had never forgiven him for his pressure upon them in the early days of sound movies.

After the great signatory pact in 1928 that licensed most of the motion picture industry to make sound movies only according to ERPI's terms, competition should have been stifled because of those powerful clauses in both producers' and exhibitors' contracts, according to the Bell point of view. But it was not, and so Mr. Otterson and William J. Fox had conceived a plan whereby ERPI would literally clear the field by buying up the movie empire and making it no more than a Bell-Fox agency. So, upon Otterson's recommendation, and without any apparent trepidation, the Bell system loaned William J. Fox $15,000,000 with which to purchase whatever stocks he saw fit for development of control over the movie industry of America. The loan was made in February of 1929, on a note of one year's term. Up to the time the money was passed, relations between Fox and Otterson were exceedingly cordial.[8] Mr. Otterson sprang to the task of organizing a staff of salesmen and technicians to install the sound equipment in studios and theaters. Nor was Mr. Fox idle. He dashed into the markets with his borrowed $15,000,000 and began a vigorous attack upon the stock of Loew's, Inc., which owned and controlled a vast chain of theaters in addition to the producing company, Metro-Goldwyn-Mayer. He went to Europe in grand style, and there bought up a little flywheel device that was destined to excite joy and music for him and give the great Bell system many a worried hour, until the Supreme Court of the United States, with its mystical understanding of the Constitution and the patent laws, dissolved all fears for the corporation and sent Mr. Fox into a torment of frustration.

That flywheel arrangement was known as the "Tri-Ergon" device for governing the operation of motion picture sound film recording cameras and exhibition machines. Remember it well.

The record shows that Fox, still playing the game of smiles and soft words with Otterson, suggested that if his good friend at ERPI put up half the money, he, Fox, would put up the other half, to acquire this little instrument at the nominal price of forty thousand dollars.[9]

For some reason, ERPI failed to heed the old precepts of Gardiner Hubbard, the tried and true Bell policy of buying up competition. Otterson, in lordly fashion, turned down the Tri-Ergon proposition, and William J. Fox went into it alone. What he did not tell Otterson at the moment was that he bought the Tri-Ergon patents privately and personally as William J. Fox, not in the name of the Fox Theatre Corporation, to which the Bell system had lent the $15,-000,000, and with which the Bell System had a reciprocal licensing agreement so that all developments of one corporation could be used by the other.[10] On account of that little secret of Mr. Fox, trouble began to brew in Eden.

There arose an exchange of less and less warm correspondence between the two friends after the Bell patent lawyers and scientists told Otterson what they appeared to have let slip past them. For at that time it appeared that Tri-Ergon technique was the only successful method of making sound movies—that the device literally represented a ruling patent. By coincidence, just at the time the ERPI high command began to wonder how it was going to recover control of Fox and his Tri-Ergon device there came a market crash. Fox appears to have felt, and has so implied in print, that the crash was just a trick to get the Tri-Ergon patent out of his hands and into Otterson's. Other people would be disposed to point out more valid reasons for the collapse, but anyhow Fox's story makes him the hero, and that's the natural inclination of any man, movie magnate or furrier.

Fox Theatres fell dizzily on the market. Came February 26, 1930, and a no longer amiable Otterson demanded $15,-000,000 in cash and on the dot. Long since, William J. Fox had been skidded out of control through the operations of a trust agreement he had made on December 3, 1929, with Charles Evans Hughes, the present Chief Justice of the Supreme Court, as his counsel.

The trustees of that agreement had been Otterson, Harry Stuart of Halsey, Stuart, and Company, stock brokers, and Fox. But Fox still had his Tri-Ergon claim, and he pressed it. He brought suit against ERPI for patent infringement and demanded an accounting of profits that would run into hundreds of millions of dollars. He lost the case in the U. S. District Court in New York, won in a review by the U. S. Circuit Court of Appeals, and presumably won again from the final arbiter, the Supreme Court.

It appeared that the Bell system would be forced to pay, to the man whose theatrical career it had liquidated, estimated damages in the amount of $200,000,000 cash, and further royalties upon every film produced through use of the Tri-Ergon device.

Here was a furrier's triumph—or so Fox thought. But he reckoned without the resources of the Bell system. The Supreme Court decided to rehear the Tri-Ergon Case. Such a determination to consider for a second time a matter which has been fully adjudicated is rare in the annals of the Court.

On rehearing, the Court concluded it had been wrong about the Tri-Ergon patent. It then decided that the claim was not valid, and that Fox had nothing coming to him.[11] A sadly disappointed Fox went slinking off to retirement. Some mighty minds were eased, mightily, for the moment, but the sound motion picture problem still plagued them.

RCA, with its Photophone and its RKO pictures, was still giving competition of a sort. The Bell system does not rest easy so long as one competitor is in the field, perhaps with a chance to grow. And so, in April, 1932, ERPI set up a revolving credit of a half million dollars with which to finance production of motion pictures and to operate studios as an offset to RCA. By June, 1933, the revolving credit was inflated to $800,000 and the staid Bell system was right in the middle of movie making, with girls, comics, villains, and grease paint all around.[12] Some of its productions were the remarkably unsuccessful offspring of Ben Hecht and Charles MacArthur, such as *Crime Without Passion, Once in a Blue Moon,* and *The Scoundrel.* Others, like *Moonlight and Pretzels,* were mildly fruitful at the box office.[13]

In all, ERPI was involved in the production of more than a hundred minor features, short subjects, comedies, and industrial films of the sort that a small competitor might bring out and, by luck or other odd chance, make a fortune on with which to develop into a formidable opponent of the Bell system for the cash of Hollywood.

Otterson's epilogue to the direct production operations of ERPI among the little independents was:

> The successful operation of this studio has driven practically all of the bootleggers in the East out of business and also the studios licensed by RCA. RCA formerly had four such studios which are not now operating. . . .
>
> Through our financing of pictures we have gotten a steadily increasing proportion of the business and have left RCA with little or no income from royalties except in connection with studios owned and operated by themselves.[14]

Who is this RCA that worries the Bell system so? What can it do, either in telephony or television, comparable to the threat it offered in the wholly nontelephonic field of sound motion picture operations? The Bell system never overlooks a competitor. Television appears to be a business that cannot function except in monopoly. Let us examine the Radio Corporation of America.

16. RCA Pays a Dividend

WOODROW WILSON, AT VERSAILLES, WAS A MUCH SOUGHT after man, but nobody pursued him more ardently than some admirals of the United States Navy, concerned for the future of radio in the Western hemisphere.

Just what passed between him and them has never been brought into the public record altogether. Death, partisan passion, and imperfect memories have all helped to obscure the details. But at any rate the admirals caused him to take some sort of action. And because of his action, two letters of extreme importance to the people of the United States were exchanged. They are quoted here in full, and their historic value will be apparent to every reader:

Hon. Franklin D. Roosevelt,   March 29, 1919
    Assistant Secretary of the Navy,
        Washington, D. C.

Dear Mr. Secretary:

In view of Admiral Griffin's absence from the country and of the pressing importance of the situation to which I refer in this letter I am taking the liberty of writing you in regard to a letter which the General Electric Co. received from him, dated February 25, reading as follows:

"The bureau requests the professional services of your research engineer, Mr. E. F. Alexanderson, to visit the naval radio station, Sayville, Long Island, to make a report on a speed control system for the high frequency alternator installed at that station.

"It is requested that you advise the bureau of the probable date of Mr. Alexanderson's visit to the station, so the necessary arrangements can be made for his visit."

As I think you fully appreciate, it is the hearty desire of the General Electric Co. to co-operate with the Government in its undertakings in every practical way, and we believe it is not the desire of the Navy Department to request us to do anything which would be inimical to our commercial interest.

We have, over an extended period, been negotiating with the Navy Department in regard to furnishing several of our radio devices, including a photographic receiver, barrage receiver, and methods of the simultaneous sending and receiving of radio messages and we now have a contract with the Navy Department for completing the installation of the New Brunswick high-power radio station.

At the same time, we are in active negotiations with the British and American Marconi Cos. for the sale to them of a substantial number of our high-power radio equipments with the necessary accessories, of which the above mentioned devices are a part, including a license to those two companies to utilize our system commercially on a royalty basis.

In view of the foregoing circumstances, I think it would be extremely helpful if we could in the immediate future have an opportunity to talk this situation over fully, for the purpose of arriving at a mutually satisfactory understanding whereby we would be in a position to furnish such equipment and such engineering advice to the Navy Department as may be required from us, and at the same time retain a reasonable protection of the commercial interests of the General Electric Co.

If this suggestion meets with your approval and you will

kindly name me an appointment, I shall be pleased to go to Washington with other representatives of the General Electric Co. to discuss this matter with you and others of the Navy personnel who are immediately interested.

Very truly yours,
Owen D. Young.[1]

And:

Navy Department,
Washington, April 4, 1919

Sir:

The Department appreciates the spirit of your letter of March 29, dealing with the purchase by the Government of your numerous excellent devices for radio-telegraphy and your pending negotiations with the British and American Marconi Cos. Due to the various ramifications of this subject, it is requested that before reaching any final agreement with the Marconi Cos., you confer with the representatives of this department. It will be greatly appreciated if you and other members of your company call at the Navy Department to discuss this matter at 10 A.M., Friday, April the 11th.

Very respectfully,
Franklin D. Roosevelt,
Assistant Secretary of the Navy.[2]

Mr. Owen D. Young,
Vice-President General Electric Co.,
120 Broadway, New York, N. Y.

Just why were the admirals bothering their President? Why was the Assistant Secretary of the Navy so anxious to keep those "excellent devices" handy for the Navy's use? To understand, one must consider the situation in the world just before the Great War opened. At that time wire cables were the links between nations, and the cable companies were all dominated by Great Britain. No nation's diplomats felt they had any secrets from England, no nation could carry on business sure of privacy. That sort of thing continued to obtain even after the development of radio, as in the instance when Baron Aloisi of Italy was cut off at London while attempting to tell Il Duce's version of the Italian aggression in Ethiopia to the American audience by way of a circuit that had to be rebroadcast from England.

But in 1919 the United States, with a powerful and modern navy, with a huge store of war supplies and several million men mobilized, was poised for dominance of world diplomacy. The Navy had no intention of letting anything so important as independent international communications escape. It learned that the Marconi Wireless Telegraph Company of America, really British controlled, was negotiating for rights to exclusive use and sale of a "very excellent device" called the "Alexanderson alternator," which had been invented by Dr. E. F. W. Alexanderson, of the General Electric laboratories. In 1919 the Alexanderson alternator would furnish a high frequency current vastly more efficient than any other available for radio, and the nation controlling its sales might easily condition the growth of the new industry. And here was General Electric, as Mr. Young's letter shows, about to sell the Alexanderson alternator to the British controlled Marconi company. Had the sale gone through, there is no way of telling what might be the status of radio today. Britain has never been openhanded in her favors to other nations.

It would be interesting to know the details of Mr.

Young's conference with Mr. Roosevelt. General Electric's position was simple. It was in business to serve customers as they came. Great Britain was begging to be a customer. And the Navy Department apparently was hesitating. But the Navy had some weapons of its own. The Government had seized all radio during the war, and the Secretary of the Navy was both vehement and vocal in favor of a peacetime policy of complete government ownership and operation of all radio message services.[3] Nobody even considered the possibilities of radio as a broadcaster of entertainment, apparently. The government was in a strong technical position to bargain, if not to control; for the Navy Department then owned a great number of unadjudicated patents on radio, some of which had been seized from Germany by the Alien Property Custodian. At least one of these appeared so good that, in competition with one of Mr. Alexanderson's inventions, it was declared the controlling patent in a test throughout the Canadian Courts and lost only before the Privy Council of Great Britain.[4] Finally, the Navy Department could always use moral suasion on grounds of patriotism.

The world has never been told just how it was accomplished, but at any rate General Electric compromised and bought out the America Marconi Company, and on October 17, 1919, set up the wholly American owned Radio Corporation of America, to sell wireless equipment. Shortly afterward a sensational story broke out in England concerning dealings in the Marconi holdings,[5] and a strong governmental influence was made apparent in the organization of RCA. Major General James G. Harbord (retired) became president and Rear Admiral W. H. G. Bullard, chief of naval communications, sat officially at the meetings of the board of directors as a representative of the national defense system. And the government did radio business almost exclusively with RCA, even going so far as to abandon use of its own patents as weapons which might bring down the price of equipment, or open up the field for competition.[6] This was a conscious, official policy, so described years later by a naval officer in authority during a congressional inquiry.

There is a curious, persistent pattern of collateral human history tied in with the history of radio. When RCA was organized, one David Sarnoff was made its commercial manager, the same David Sarnoff who, as a wireless telegrapher for the Marconi company, had been shrewd enough to sell that historic message concerning the sinking of the *Titanic*.

By 1937 David Sarnoff was a powerful world figure as President of RCA. Ex-Assistant Secretary of the Navy Franklin D. Roosevelt had also become world powerful in 1937 with the undoubted aid of radio, which had served him vitally in two successful campaigns for the Presidency of the nation. It had served RCA and Mr. Sarnoff, too.

The RCA was, by that year, maintaining radio communication between the United States and forty-five other countries. It offered ship-to-shore communication, photoradio (facsimile) transmission service, photophone sound equipment for motion picture theaters and producers, national and international radio entertainment (National Broadcasting Company), laboratory research for licensee manufacturers, and numerous small subsidiary services involving sight and sound. It maintained an institute for training radio engineers and publishing reports on its laboratory research. It either manufactured directly or licensed all sorts of sound and sight radio equipment, sound motion picture instruments, and phonograph equipment and records.

In thermionic valves for sound radio receiving sets its position once was declared illegally monopolistic in a court of law. This will be more fully treated later. Its net income for 1936 was $6,155,937, and, though it had started its career in the fond embrace of the Federal Government, had always been recognized as dominant in the radio field, and had seen that industry become so powerful in American commerce as to involve the spending of $900,000,000 in that year [7]—in spite of all these things, by 1936 RCA was still so uncertain of its future that it had never paid a dividend. In 1937 RCA startled the communications industry by declaring a virgin payment to its stockholders (approximately two hundred and fifty thousand) of twenty cents per share, giving fresh impetus to the rumor that "Rockefeller's in" with a mission to clean up all confusion in RCA's corporate structure before the advent of television. But RCA stockholders, by then, were veterans at being startled. They had seen the quoted price of their shares rise on the New York Stock Exchange from $1.50 to $549 each. They had also felt the effect of stock market pools upon their equities. One of those pools was famous for the list of distinguished participants—among them the well known editor Herbert Bayard Swope, who put up no cash collateral but received a profit of $58,342.15. Others who profited without risking cash included T. J. Ward ($87,513.24), J. J. Riordan ($58,342.15), and Mrs. M. J. Meehan ($87,513.24).[8]

But with all its wild Indian behavior on the New York Stock Exchange, its sweet benefits to short sellers and disappointments to simple seekers after dividends, RCA is far from a puny threat to the great, solidly financed and ultra-respectable Bell system. Its stocks have behaved wildly, but so has the electron. The years since 1920 have been as turbulent in the laboratory as they have on the Exchange. And the Bell system has kept its balance up to now by making treaties, as solemn and as vital to its position as ever were any compacts between sovereign nations—and, as we have remarked before, about as rigidly kept. Corporations respond to technological change as inevitably as do governments to rising birthrates and declining prosperity.

The Bell system, at about the time Mr. Young and Mr. Roosevelt were thick in their negotiations, was coming to the horrified conclusion that radio telephony might some day be conducted on a two way basis, just like wire telephony. What would then become of the great Bell system? The air, in 1920, was truly electric. New words were creeping into the common language, new conceptions of time and space. Amateurs at radio communication called themselves "hams" and spoke wisely of "pickle tubes" and "cats' whiskers" and "crystal detectors," and a few department stores with a flair for novelties were offering receiving sets for sale.

And the Bell system, always mindful of that trading philosophy which demands even yet that it protect its monopoly on domestic telephony at all costs, was drawn deeply into the growing business of radio invention, communication, and manufacturing by the compulsion of threatened competition.

Consequently on July 1, 1920, it signed a set of stipulations with General Electric, RCA, and several other companies which came to be known as "the radio group."[9]

Technically this set of stipulations was called a "cross-licensing agreement," the first of a series forced upon both the telephone group and the radio group by the progress of invention. It might be added that the last such agreement has not yet been made. The announced purpose of the solemn business treaty was to break a deadlock in patents and allow the useful art to advance. There was certainly some ground for such a view. The telephone group held certain vital patents, such as that governing the De Forest Audion. The radio group, with the Alexanderson patents, was equally powerful in basic equipment control. And the United States Navy, still clamoring for government ownership or control of radio, was demanding that the two groups exchange information in the national interest. And so they did, but the agreement did not end with simple exchange. Restrictions were put into the use of every patent, and those restrictions had ramifications almost as infinite as those inherent in the operations of the radio spectrum. They have undoubtedly changed the course of invention and corporate history in electronic communications. To accomplish this, some marvelously intricate patterns of behavior were laid down.

For example, the Bell system assigned all its patents to the radio group, but restricted the radio group from using for competitive purposes any patents involving telephony, either by wire or wireless—perhaps an innocent seeming notion when first it confronts you. In turn, the radio group handed over to the Bell system all its own patents, but with the restriction that none could be used for a competitive radio message service. Now the importance of the agreement becomes obvious.

As the business of services was restricted, so was that of

manufacture. In general, the Bell system was allowed broad leeway in the making of transmitting equipment for all types of radio. The radio group, especially RCA, leaned toward exclusiveness in the making of receiving equipment. It leaned so effectively that nearly every radio receiver in the United States today is produced under its patent, and every purchaser pays royalties to it in addition to the cost of the set fixed for the benefit of the ultimate retail merchant.

But the 1920 treaty, first of its kind, had serious imperfections. Broadcasting of entertainment was far more tenuous and vague then in the public mind than television is now. Failure to be explicit in assigning rights and usages in connection with broadcasting led, within a year, to violence and undeclared wars between the treaty signatories; for just four months after the signing of the compact there occurred an epochal happening. The Westinghouse Electric and Manufacturing Company, a vigorous experimenter (and not a signatory to the original treaty), had on its staff an engineer, Dr. Frank Conrad, who operated a broadcasting station in the garage back of his home at Pittsburgh, Pa. Dr. Conrad's station had a limited but enthusiastic following because he made a practice of sending out interesting programs. On election night, 1920, he made news that caused the more farsighted publishers of daily journals to shiver with apprehension. He broadcast the details of Warren G. Harding's victory in the Presidential campaign. Conrad's station became famous as KDKA, Pittsburgh; and the signatories of the 1920 agreement found Westinghouse riding on the crest of a wave of favorable publicity, a competitor which must be brought into the treaty.

And the treaty, under the strain of popular demand for

more and more and more of radio, collapsed even though Westinghouse bound itself to the same terms governing the rest of the radio group. The sensational performances of KDKA made it obvious that a radio station owner would be a king in his community. And it was equally obvious that a linking of several stations together for a simultaneous broadcast would make an even more resplendent emperor of him who controlled the chain. This situation obtained at a time when the Federal Government appeared to be powerless to withhold or withdraw a license from any applicant. With radio becoming a great power in human affairs, what would the Bell system do about it? In 1923, the question was pressing. A conference of management officials was called in New York City, with A. H. Griswold, vice-president of the A. T. & T. in charge of radio matters, stating the proposition thus:

> We have been very careful, up to the present time, not to state to the public in any way, through the press or in any of our talks that the Bell system desires to monopolize broadcasting, but the fact remains that it is a telephone job, that we are telephone people, that we can do it better than anybody else, and it seems to me that the clear, logical, conclusion that must be reached is that sooner or later in one form or another, we have got to do this job. . . .[10]

Griswold proposed, that, in order to do the job properly, the Bell system organize radio stations in every possible community, with "representative citizens" in charge of programs but with the Bell system building, operating, and owning the stations and receiving sets therefor; quite a typical, tight, little Bell monopoly indeed—and as we look back upon it, possibly the best solution to what has become an intolerable conflict of engineering theories on methods of broadcasting. The Bell group, whatever else one might say about them, could be depended upon to give uniformly good service.

Some portion of the Griswold program actually was achieved before external forces bogged it down, but it was clear, almost from the start, that the Bell system and the radio group were bound to have further conflicts, regardless of the treaty. The tremendous public demand for equipment, the novelty of both the art and the operations of radio, constituted a pressure too great to be withstood. The Bell system opened its offensive-defense with a powerful station at New York, WEAF, and began to acquire others in such strategic cities as Washington and Chicago.

By the time of the conference of 1923, lines of conflict in the sales of equipment and the operation of stations were drawn and battle was imminent. As a result of the Griswold conference, the Bell system adopted a technique of handling opposition which might be considered a worthy piece of evidence of what it may do about television. It refused to offer its wire network as a public service responsibility and to assist development of general radio broadcasting. However, in specific cases it did allow the open use of its wires for radio when no conflict requiring expansion or surrender of regular telephone operations was involved. Furthermore, special exceptions in which wires would be granted instantly were (1) for stations owned and operated by the Bell system, (2) stations belonging to the Government, and (3) for stations licensed under Bell patents, providing in each instance approval was obtained in advance from Griswold.

Griswold warned the associated Bell companies that if they were to provide wire telephony as an adjunct to radio broadcasting stations not licensed under Bell patents, they would tend to jeopardize certain important patents under which the companies themselves were licensed and operating.[11]

The Bell system, obviously, had better wire facilities for broadcasters than any competitor. Also it was offering, through Western Electric, good speech input devices and other instruments of broadcasting. But unless a station were fully equipped by Bell, it could not get service. That was the same fashion of doing business that Otterson used, but with a certain lack of finesse, in the sound motion picture business of 1927–1934. Griswold's policy of "whole hog or none" appeared to be working around the radio treaty of 1920. Early in 1923, the Bell organization was moving very smoothly toward domination of the new art, even though the treaty appeared superficially to have transferred control to the "radio group" which by then included Westinghouse.

But the radio group was resourceful. It demanded that the Bell system give service to non-Bell stations held within the radio group. When the telephone organization hesitated, the radio men threatened to license Western Union and Postal Telegraph to use Bell telephony patents encompassed by the 1920 treaty for development of an adequate network of radio wires. A critical legal-technical question arose. Are wires incident to broadcasting and simply a part of radio; or are they separate so that radio programs are to telephone wires no more than ordinary party calls? Naturally, if wires were to be considered an adjunct of radio, then the Bell patents could be used to build a great national network, wholly independent of the Bell system, for radio use. Here was a terrible threat to monopoly, for a standing wire network might be converted to any use. It might develop telegraphy to a point where telegraphy could recover a competitive standing against the telephone, or it might be converted by governmental order into a parallel telephone service.

The champions agreed to arbitrate. A Boston attorney, Roland W. Boyden, was selected as referee, and the telephone and radio groups submitted their arguments to him in short order, for time was valuable. Briefly, the position of the telephone group was that it had full power to force radio stations to use its equipment exclusively by refusing service to non-Bell licensees; that the use of its wires in broadcasting was an incident to telephony, not to radio; and that the radio group had no right to license the telegraph companies to use Bell patents as an aid to radio transmission.[12]

Of course the radio group argued the exact opposite; and Boyden, in a decision given on November 13, 1924, agreed with the radio group. Wires were incidental to broadcasting, he held, and so the radio companies were empowered to grant nonexclusive licenses to anybody they chose to set up broadcasting facilities for their stations. But it was not all so simple, the radio men decided, in reflecting upon their victory.

The Bell system, remember, was very great and powerful in 1924, and its laboratories were working overtime. Nobody knew what it might bring out next, or how its banking supporters might choose to retaliate upon injury to their favorite stock. And, in fact, the subsequent conduct of the Bell organization in the sound motion picture field, we already know, was such that a competitor might think long about offending it too deeply. Nothing could arouse the telephone men more (and the radio group knew it) than to license Western Union and Postal Telegraph to use Bell patents for the construction of great national wire circuits for broadcasting—especially since those circuits might easily become the basis of a governmental "yardstick" telephone system. Victory by such means might bring as great peril to the radio group in Wall Street as to the Bell system in Washington.

Consequently on July 1, 1926, a new treaty [13] was negotiated, at about the time when the Federal Government's efforts to regulate radio were being blown sky high in the courts and piracy of frequencies was common. The 1926 treaty was considerably more complex and more detailed than that of 1920. It represented, in a way, the infinitely ramified conflicts of interests which had grown with the growth of the electron's uses. One fact appears definite; that the agreements between the contestants did far more than any act of the Government to clarify the radio situation of 1926–1927, and to make for orderly expansion of radio usages, even if at severe cost to those unfortunate entrepreneurs who were not on friendly terms with the great powers.

17. The Trust Dissolved?

THE TREATY OF 1926 WAS IN THREE PARTS. FOR ONE MILLION dollars, the Bell system transferred to the Radio Corporation of America its license for station WEAF in New York, and withdrew entirely from competition in broadcasting of programs. In return, RCA agreed to use Bell wires exclusively, regardless of the cries and cut rates of Western Union and Postal. RCA agreed not to compete with the telephone company for telephone business, and in return received important rights for exclusive manufacture of receiving sets, a rich business which the Bell system otherwise might have jeopardized by competition, either directly or through a license to other companies.[1] In effect, the telephone company handed over the field of radio, except for transmission equipment and transmission length by wire, and said to RCA, "Go ahead and settle the competition any way you like. Just give us the business in transmission length."

The treaty was a masterpiece, right enough. Only one thing was wrong with it. The Federal Government, under the insistent pressure of would-be competitors of RCA, came to the conclusion that the treaty was unlawful conspiracy in restraint and monopoly of interstate commerce;

and it filed suit, under the Sherman Anti-Trust laws, on May 13, 1930.[2]

Now there are some curious undercurrents of sentiment and some remarkably brilliant decisions to be found within the workings of the Bell system's administrative bureaucracy. The true Bell man is a telephone man, pure and simple. Just as he never liked the great sound motion picture uproar within ERPI, involving Mr. Otterson, Mr. Fox, et al., so was he reluctant to tie telephony too closely to Mr. Sarnoff and the rest of the radio group. At the outset of the formation of radio policy each of the principal interests in the field—General Electric, Westinghouse, and the A. T. & T. Company—had taken some stock. In 1926, after the formation of the treaty, RCA had organized a program service company called the National Broadcasting Company.

But the Bell system, with traditional foresight, liquidated its RCA holdings in 1923. Therefore, in 1930 it was clean and clear of any financial connections with the opposite signatories of its treaty. Bell simply furnished service under an exclusive agreement, and offered complete equipment for broadcasting. So does it, today. And when the great controversy concerning the "radio trust" was settled, the Bell system was absolved for everything except its exclusive broadcasting service contracts. In 1930, when the Government's anti-trust action was brought, the gross income to the telephone system for radio wire was $4,410,904.73. The gross income of Western Union from that source in the same year was $6489, and Postal reported $3133. In 1935, the Bell system's gross receipts under the service agreement amounted to $4,529,162.57, while Western Union drew a gross of $10,754, and Postal $18,865. On a routine

THE TRUST DISSOLVED? 179

day (March 31, 1936), the Bell system had 24,949 circuit miles of wires set up for the National Broadcasting Company, and 17,217 circuit miles for the Columbia Broadcasting System. How many more thousands of circuit miles it had operating that day between the hundreds of independent stations, nobody has attempted to estimate. On another routine day (July 29, 1936), Western Union had just a little less than three hundred circuit miles in service, and on August 6, 1936, Postal was operating 3369 circuit miles for radio.[3] The Bell system has preserved itself against competition in sound radio, as it has in sound motion picture.

But what about the Government and the "radio pool"? What ever became of the Radio Corporation of America and its original policies of strict governmental interest in equipment sales? The respondents to the Government's suit of 1930 were: the American Telephone and Telegraph Company and its subsidiary, the Western Electric Company; the Radio Corporation of America from which the A. T. & T. had extracted its property interest; the General Electric Company, the Westinghouse Electric and Manufacturing Company, RCA Phototone, Inc., RCA Radiotron Company, Inc., RCA Victor Company, General Motors Radio Corporation, and the General Motors Corporation. All these might properly be called just the corporate victims of an ex-newspaper correspondent, Oswald Schuette. Certainly the anti-trust action against them was nothing less than Mr. Schuette's personal victory, a victory which has been compared to David's over Goliath.

It was a formidable bill of complaints that Mr. Schuette and the Government drew up against these towering giants of industry; and the trouble centered, of course, upon the cross-licensing arrangements. The practical effect of the 1926 treaty, the Government alleged, was to prevent competition with the telephone company for wire or wireless telephone or telegraph service or equipment in the United States, and to debar competition with RCA for similar business between nations. The telephone company actually debarred itself from using its own patents or licensing others to use them in the radio industry to compete with RCA, and RCA debarred itself from using its own patents or licensing others to use them in competition for point-to-point communication in the United States.

Mr. Schuette's interest was directed most emphatically toward the matter of licensing competition for radio sets. Let us see why. The story is simple and, as they say in the movies, heart-rending, for it is the short and simple annal of the independent entrepreneur liquidated, like the Russian Kulak.

As order, in a relative sense, came out of the chaos of the 1926 breakdown of law, radio stations sprang up all over the country in fierce competition. That broadcast of 1920 from Dr. Conrad's garage had set inventors to tinkering at a pace probably never exceeded in history. Manufacturers blossomed like the flowers of spring, just as gay, and just as sure to wilt. And wilt many a one did in short order. In 1923, the House of Representatives adopted a resolution demanding that the Federal Trade Commission investigate the "common assertion that the development of the art, its use and enjoyment, is being hampered and restricted," by closely affiliated interests seeking a monopoly.

The Trade Commission was back before Congress within seven months to report that the members of the radio

THE TRUST DISSOLVED? 181

group had conspired to monopolize manufacture and sale of equipment, and the service of communication as well.[4] For four years thereafter the Trade Commission went about the business of collecting evidence upon which to base an action against the radio combine, but in 1927, just as it was getting set for trial, the Supreme Court of the United States ruled that it had no power to act against violators of the anti-trust laws. Such anti-trust suits were a prerogative of the Department of Justice, the Court held, and if the Justice Department would not act, then nobody else could.

RCA's position was very simply put by Col. Manton Davis, its general counsel:

> There has been recently an amelioration of that policy [of withholding licenses from would-be competitors], with respect to press associations that, having received licenses from the Federal Radio Commission, desire to establish communication services.
> As I understand the expressed policy of the company, there is no other amelioration of the policy of the company to decline to furnish either the swords or the guns by which other people can enter the fields in which it operates.
> Those devices, as we have frankly pointed out, are covered by patents, and our conception is that we have a right to sell or not to sell, to sell for a good reason or for a bad reason, or for no reason, and not to sell for a good reason, a bad reason, or no reason.[5]

That is a simple, forthright statement of views. It is the traditional position of the vigorous business man, organizing his resources and standing on his legal rights to use what is his however he sees fit, within the law. Ah, yes, within the law But what was the law governing RCA?

The radio act of 1927 stated that the licensing authority (the Federal Radio Commission) was directed to refuse a station license or the permit for construction of a station to any person, firm, company, or corporation, or any subsidiary thereof finally adjudged guilty by a Federal court of unlawfully monopolizing or attempting unlawfully to monopolize radio communication, directly or indirectly, through the control of the manufacture or sale of radio apparatus, through exclusive traffic agreements or by any other means, or of having used unfair methods of competition.[6]

This was a potent bit of legalism. Remember what it provided. The same principle has been extended, incidentally, into the currently controlling Federal Communications Act of 1934. And it ought to have been enough to give pause to Colonel Davis and the policy-making officials of RCA, who had a very definite stake in radio broadcasting licenses by virtue of the National Broadcasting Company and other subsidiaries in both domestic and international radio communications.

RCA, all this notwithstanding, continued a policy of granting licenses to some set makers and sellers, and refusing them to others, "for good reason, bad reason or no reason," with the result that the few who kept both their nerve and their solvency ultimately set to work not only Mr. Schuette, but also the committees of Congress, the Department of Justice, and the Federal courts. They besieged the Federal Radio Commission, demanding that it invoke the anti-monopolistic provisions of the radio law. And they threatened to take up the issue with the voting public. Indeed, some did as much.

The Constitution grants a patent holder great latitude in

his own use of his own patent. However, the commission had a mandate to withdraw RCA's licenses of station operation if it should be found using patents in monopolistic or unfair manner. But nothing really saved the independents. Today, after all their efforts, they are, as a class, only "independent licensees" of RCA, their corporate lives simply paper grants of existence.

The Presidential campaign between Alfred E. Smith and Herbert Hoover—who, as Secretary of Commerce, had posed for a television broadcast and reshuffled American radio just the year before, allowing it to become the commercial enterprise it is today—took public attention away from the intra-industrial squabble when it exploded in 1928. In the great prosperity year of 1929 only placid smiles answered resolutions in House and Senate that the Department of Justice investigate the Federal Trade Commission's report on the radio industry, but the Trade Commission saw a storm brewing and was prompt to relieve itself of further responsibility. It loaded trucks with the ten thousand pages of testimony and evidence, stacked up an armload of report copies, and shipped them over to the Attorney General of the United States. Much newspaper space was devoted to the fulminations of Mr. Schuette. And still nothing happened. The Attorney General just whittled. Finally Congress became enraged at departmental indifference; apparently nothing at all was to be expected from the Federal Radio Commission in spite of the congressional law that ordered it to withdraw licenses from stations of companies that misbehaved. And so a resolution was adopted authorizing the Senate Committee on Interstate Commerce to "provide for the regulation of the transmission of intelligence by wire or wireless." [7] This brought the lobbyists out of their placidity. The Radio Corporation of America decided to fight Mr. Schuette on his own ground. Owen D. Young granted an interview to the *Saturday Evening Post* purporting to show that RCA had been founded at President Wilson's especial request. Nothing was said about the Navy Department's expressed policy of refraining from use of government owned patents as a means of stimulating competition. Then a full page advertisement in the *New York Times* stated the RCA plea of good intentions so concisely that we give it here in full:

THE RESPONSIBILITY FOR LEADERSHIP IN RADIO

A message to the Radio Public:

The responsibility of leadership in radio rests squarely upon the shoulders of RCA, because as the creator of broadcasting science [no mention of Westinghouse's Dr. Conrad], the creator of broadcasting apparatus [no mention of Marconi, Fessenden, or the other pioneers], as the creator of dependable transoceanic wireless [The Bell system had broadcast from Arlington, Virginia, to Paris in 1915], RCA made it possible for the public to have broadcasting! [Exclamation point ours.]

RCA, founded at the request of the United States Government before our troops were demobilized, was expected to blaze the way in the radio field,—scientifically, commercially, patriotically.

This was a very clever story. It appeared on October 2, 1929, at just about the time when Mr. William Fox was getting his bad news from the Bell system's Mr. Otterson that unless Fox were to hand over the Tri-Ergon patents to ERPI something unpleasant would be bound to happen.

But clever as it was, the RCA's campaign to capture public good will failed under the pressure of events. For, it appears, the stock market crashed. With it, as we have seen, crashed William Fox's dream of empire. And with it crashed the dizzily soaring Exchange price on RCA's non-dividend-paying stock. And with that, a lot of public good will. Government, with its usual weakness for a good devil hunt, became acutely interested in demonstrating its zeal for the public good by looking for somebody with a lot of money to prosecute. Suddenly, as the chill of winter settled down, the hearings before the Senate committee became vital proceedings. Copies of the record became scarce. Today they are rare editions of Washingtoniana.

With 1930 events took a turn for the worse. Lee De Forest (how curiously these human fates weave in and out of the story of the technology) had decided that he, of all men, ought to make a fortune out of radio. With his genius and the profits from assignment to the telephone company of his original patents for the thermionic valve, he went into the radio business.

His was a magic name in the industry. Why should it not be with the public? The De Forest Radio Company offered equipment that should have brought the aging inventor a fortune. Or so he felt, in his naïve conception of the business man's career. But RCA had no intention of allowing De Forest or anybody else to invade its field. It sold its own thermionic valves to contractors in radio equipment (and they were good valves) only on the condition that the purchasers would refuse to accept the equipment of any competitor of RCA. That, in effect, was the application to the radio equipment business of the same tactic that the Bell system's Mr. Otterson used in his original ERPI contracts controlling sound motion picture equipment. Otterson had used it supremely well to limit RCA's competition in that industry, but RCA was less skillful in its methods. Each committed costly blunders. The great difference between the RCA-ERPI contest and the RCA-De Forest case was that in the one instance the seemingly weaker power had the ultimate vast resources of a great patent and financial pool (General Electric) behind it, and in the other there was no buttress except the abstract law and the inventor's good name. The result was bankruptcy for Lee De Forest and in 1930 the fruits of policy were served in the form of a suit by Arthur D. Lord, receiver for De Forest, charging RCA with violation of the Clayton Act of 1914 which prohibits unfair trade practices tending to create monopoly and restrain trade.[8]

And most cataclysmic of all, the Department of Justice was finally moved to bring suit against all signatories to the treaty of 1926, charging them with violating the anti-trust laws. The long campaign by Mr. Schuette was having its effect at last.

How the RCA escaped destruction in this era of converging misfortunes is an untold miracle. On the one hand it was tied to the Bell telephone system by a treaty governing radio and wire transmission of intelligence and operating so satisfactorily in that field that the Federal Government found it necessary to bring action to dissolve the agreement. On the other hand, in the business of sound motion pictures, RCA was fighting the Bell system tooth and nail for the right to compete. Neither would agree to bring that profitable side venture within the scope of the treaty. And here was RCA, using its pool of patents under that same treaty, to license or refuse to license outsiders in the radio industry, "for good reason, for bad reason, or for no reason," in such a fashion that the Federal Government, this time in the role of a bankruptcy referee, was suing it for unfair trade practices. This was an extremely involved and precarious position for any company to be in at any time, you will agree; and it was made even more complex and difficult because of the general condition of business in 1930. But RCA was equal to the problem.

Both the De Forest bankruptcy case and the Department of Justice's anti-trust suit were brought in the United States District Court at Wilmington, Delaware. Let us see first what happened in the De Forest case.

The Court handed down a decree in which the Radio Corporation of America was adjudged guilty of unfair trade practices "to substantially lessen competition or to tend to create a monopoly" in the commerce in thermionic valves, without which no radio equipment, then or now, could operate. It went even further and enjoined RCA from ever using contracts again which would have such effect, making the setback overwhelming.

Here was a clear, concise statement of guilt. RCA was violating the laws of the United States concerning monopoly, and the commerce in which it was engaged happened to concern radio equipment. In the verdict there was a clear mandate to the Federal Radio Commission to give consideration to the clause in the radio act of 1927 prohibiting holders of broadcasting licenses from even so much as attempting such restraints in "radio apparatus and devices entering into or affecting interstate or foreign commerce, and to interstate or foreign radio communications."

Consideration was given. The Federal Radio Commission, by a three to two decision, found that RCA had not violated the anti-monopoly provisions of the radio act.[9] This is one of the most curious bits of legalizing ever recorded in a Federal tribunal. The line of reasoning set up by Commissioner Starbuck, with Lafount and Robinson concurring in the conclusion, is so remarkable that we give it in detail:

> As will be observed, it [the radio act of 1927] prohibits the issuance of a license or permit only where a court has found the existence of a monopoly in radio communication, (a) through the control of the manufacture or sale of radio apparatus, (b) through exclusive traffic arrangements, or (c) by any other means, or (d) to have been using unfair methods of competition. . . . Radio communication is defined in Section 13 of the Act as intelligence, etc., or a communication of any nature transferred by electrical energy from one point to another without the aid of any connecting wire. . . .
> As the decree showed, the suit pertained to a contract for the sale of goods, to wit, radio vacuum tubes [thermionic valves].
> No question of a monopoly in radio communication was involved.
> Neither the decree nor the opinions of the several courts passing upon the case contain any reference to radio communication, nor was there any finding that the contract held to violate the Clayton Act, created or tended to create a monopoly in radio communication within the meaning of Section 13 (which we have already quoted). To hold, therefore, that the foregoing decree comes within the pro-

visions of Section 13 would be to read into the Act something not therein contained. This we are not permitted to do.

The contention is made that radio receivers are essential elements of communication and inasmuch as tubes are vital to receivers, here has been such an indirect attempt at a communication monopoly as to call for the application of Section 12. To this I cannot agree.

No mention of communication is made anywhere in the various opinions of the courts or in the decree. No claim has been advanced that the tendency toward a monopoly of tubes for broadcast receivers found in the objectionable contract was of such magnitude as to stifle communication or even affect it. . . .

I am unable to conclude that receivers are such indispensable parts of communication as to preclude a monopoly thereof without the use of such receivers. . . .

Can communication be had without the use of receivers?

It would be quite possible, so far as the United States and its laws are concerned, to have a complete monopoly of radio communication to foreign countries entirely distinct from any domestic radio receivers or their tubes.

That is acceptable. . . . But that one company may use admittedly unfair trade practices to restrain trade in essential radio receiver parts and still escape the penalties of the radio act of 1927 is truly a miracle.

"Communication," says the Oxford Universal Dictionary, is the "act of imparting (especially news) information given; intercourse; common door or passage or road or rail or telegraph or other connection between places; . . ."

One may not exercise monopoly in the matter of announcing information and escape the penalties of the law, says the majority decision; but one may exercise monopoly upon an essential ingredient of devices for reception and go free. How information may be imparted without being received we do not know. This ruling is perhaps as important as any ever given in the history of radio communications. Had the commission held otherwise, it would have put RCA out of the broadcasting business and changed the whole structure of the spectrum.

Let us contrast the reasoning of the majority of the commission with that of the minority.

By Commissioner Sykes:

A careful study of this judgment and of Section 13 [of the radio act] leads me to the conclusion that this Section is applicable and that the Commission should deny these licenses. Under this Section there is no discretion whatsoever vested in this Commission. . . .

Section 15 of the Radio Act makes all laws of the United States relating to unlawful restraints and monopolies applicable to the manufacture and sale of radio apparatus and devices.

It authorizes the court in any suit, civil or criminal, in its discretion, to revoke the license of anyone found guilty of violating these laws. It is admitted by counsel for the applicant that the Delaware Court in its discretion could have revoked the license of these four subsidiary companies.

It is contended, however, that Section 13 of the Radio Act is only applicable provided the Sherman Act or the Federal Trade Commission Act has been adjudged to have been violated by final agreement.

Why should the court in Section 15 of the Radio Act make the Clayton Act applicable and omit it from Section 13?

Chairman Saltzman was even more direct:

> I dissent from the opinion of the majority of the Commission in renewing the licenses of the Radio Corporation of America. . . .
> In my opinion, the language of the Delaware District Court as hereinbefore quoted, when considered in light of the fact that vacuum tubes are an essential part of radio broadcasting receivers, and so, necessarily of radio broadcasting communication, precludes any escape from the conviction that the Radio Corporation of America was unlawfully attempting to monopolize radio broadcasting communication. . . .

David Sarnoff and his fellow workers must have felt like men retrieved from the tomb when they read the majority opinion in this case, and indeed they should have. The vote of one man saved them.

Findings of the Radio Commission were, as findings of the Communications Commission remain today, apt to be final. No external power could intervene and force the commission to deny licenses to RCA, however much the courts may have disagreed. The verdict was the commission's prerogative, granted by congressional law. Relief from a finding was provided only in the event that licenses should be denied, for then the applicant could appeal to the Federal courts. Nobody was endowed with power to appeal in the interest of the body politic to set aside a finding favorable to the licensee.

RCA's troubles were not ended with victory in the De Forest case, for there still remained the dangerous anti-trust action by the Department of Justice charging that the treaty of 1926 was a wholesale violation of law. This was to be tried before Justice John P. Nields, who has presided in some of the most important industrial lawsuits of the nation's history. Some emergency action necessarily had to be taken. Whole corps of attorneys moved into Washington and set up branches at Wilmington. Just what took place? Why was the case so long in preparation and still never brought to trial?

These are matters about which we know nothing even after exhaustive study of the public record. But this we do know. On July 1, 1932, with the special prosecutor sitting in, the telephone and radio groups undertook to amend the treaty of 1926 in such a way as to eliminate any possible grounds for charging either with illegal monopoly or violation of anti-trust laws.

In the 1926 treaty, the primary factor had been for each to assign the other patents for use in a particular field of enterprise, and to guarantee against competition. For example, the telephone company gave all its patents to RCA, but RCA agreed to use them only for radio purposes and never to foster competition against the Bell system for telephony. Conversely, RCA gave all its patents to the Bell system, which promised never to compete for radio broadcasting business or let others use RCA patents under its license for any such purpose.

These were the elements which the Government declared essentially illegal and in restraint of trade. Therefore, the signatories set out to make legal stipulations in their stead. The 1932 amendment to the treaty simply provided that each should give the other its patents, but that no company's patents could be used in competition against that company. This settled everything. For instance, the Bell system assigned all its patents to RCA, which was authorized to use them in radio any way it might choose. But if RCA should decide to compete with the Bell system in telephony, it could not use Bell patents for that purpose. Likewise, if Bell should decide to go into radio broadcasting or the manufacture of ordinary receiving sets, it was debarred from using RCA patents for such an objective.

Since, as a matter of practical fact, each company found it necessary to use the pooled patents to operate in its own field (that is, the patents of RCA and the Bell system in joint use were necessary to each in the separate fields of radio and wire telephones), it is obvious that one set of patents alone would serve neither in competition with the other. But such an agreement, however good for business, was not free of criticism. The amended treaty, before it was finally signed, was circulated among the officials of the Bell system for comment. F. B. Jewett, chief of the Bell laboratories and premier scientist of the telephone system, opposed the plan on the grounds that it was not a free interchange of nonexclusive licenses between the contracting parties. He defined it as an interchange of nonexclusive licenses with limitations of use, having the broad, practical effect of restricting the fields of possible development by each participant, even including the major activities which it was then undertaking.

Even where bilateral licenses are made, there is probably little danger of competition by the grantee in fields where the granter has already attained to a commanding position.

Thus, while a casual reading of the agreement by one not

thoroughly conversant with all the factors may appear to establish the basis for an enlarged free development in most of the fields, this is not actually the case.[10]

Jewett charged further that the cross-licensing agreements between the telephone and radio groups were stifling the science of electronics:

From the standpoint of the man who has a brilliant idea which in its first nebulous form seems applicable outside our business, there will be little or no urge to go ahead in the face of a situation where he knows that the results of his work have sold in advance outside the Bell system.

Jewett was voted down, however, and the agreement signed. G. E. Folk, general patent attorney of the A. T. & T. Company, assured the Bell system that telephony was in no way imperiled. He denied that the Bell system would be giving away monopoly rights by the agreement. If such were the case, said the matter-of-fact attorney, he could not see how he could assent to such a proposition. "Would we wish to grant to others the right, for example, to compete with us under our patents in our present field of long-distance communication, both wire and wireless telephony and telegraphy?" [11]

At another point, Folk commented:

The [Jewett] memorandum suggests "that we should use every effort to find another way out of our present difficulties, even possibly to the extent of taking the risk involved in the outcome of the anti-trust suit." The only way out that has been suggested is the formation of a patent pool—to continue until 1954—a way out suggested by the Government—and one to which we were ready to acquiesce even though it did not appeal to us.

In summary, Folk assured the Bell officials that the telephone company would lose none of its monopoly rights against radio competition with wires, the very rights which had motivated the bringing of the 1930 suit. We know, in fact, that the competition does exist, as Mr. Sarnoff has all too clearly indicated, no matter what the wishes of the lawyer may have been. It may be that Mr. Folk interpreted the meanings of the government-blessed amendment to the 1926 treaty in a way that was not clear to outsiders. It may be that the Bell system did not actually retain the monopoly rights it thought it did. But, at any rate, the new treaty was signed on July 1, 1932, in spite of Jewett's complaints, and was submitted to Justice Nields as evidence that the alleged violators of the anti-trust laws had "cleaned up the industry" of their own accord and were determined to be good, henceforth.

Whatever became of the proposal of the special prosecutor that an open pool of patents be formulated with every participant a competitor with the others, yet receiving a royalty on every use of his patent? He called it "The Electronics Foundation" when he broached the proposal.[12] A nice title. But the record shows nothing positive done concerning it when Justice Nields, with the government prosecutor consenting, undertook to review the new treaty and decide what should be done about the anti-trust suit. He was in quite an amiable mood that morning, this judge. It appeared, he decided, after looking over the compact of July 1, 1932, that the monopoly could no longer exist. The defendants no longer guaranteed not to compete with each other; in fact, they stood able to, if they chose. But they couldn't so choose (he did not add), because of the agreement that one company could not use the other's patents for competition against that company.

Justice Nields dismissed the action against the Bell system, which returned, shriven of sin, to the happy business of furnishing telephony, radio transmission wires, sound motion picture equipment, and all its ramified nontelephonic activities. But, as to the radio group, the judge entered a "consent decree" which required General Electric and Westinghouse to divest themselves of all holdings in RCA and NBC, so that RCA could manufacture radio equipment as apparently a full, uncontrolled competitor of General Electric and Westinghouse, with no eyes turned toward the "home office." Mr. Young was deposed, and Mr. Sarnoff truly became the king of the air waves; and he wasted no time in exercising his powers of both domain and diplomacy. The consent decree was handed down on November 21, 1932. Mr. Oswald Schuette, the trumpeter of the independent radio set makers, took a job with RCA as adviser of Mr. Sarnoff on matters of public policy. Dr. Jewett, still fretting about the binding influence of agreements which allowed no escape in the event of "changed conditions of laws," went back to his laboratories. Dr. Jolliffe, in good time, left the halls of government for the cloisters of industry. The independent manufacturers who couldn't get licenses from RCA to use its patents went broke, for the most part. And from the laboratories, heralding a new art and new troubles, came television, the destroyer of peace and harmony, just as Dr. Jewett had foreseen and feared. Let us see how.

The 1932 treaty defined "picture transmission" as the "art of transmitting or receiving at another point than the point of transmission," by means of electricity, magnetism or electro-magnetic waves, variations or impulses, "the aspect or shape of things, including pictures, whether still or moving, drawings, writings, forms, and other graphic, printed and written matter of all kinds and including television." [13]

And who was to get television? This takes some careful analysis. Under the 1926 treaty and the 1932 amended treaty, the general principle of division was for the Bell system to furnish speech input equipment, broadcasting equipment, and transmission length of wires; RCA, through its program service of NBC, to offer entertainment and equipment. After the sale of station WEAF the Bell system refused to offer entertainment or to make radio receiving sets for the American home. This withdrawal has had some curious practical effects. For example, in Europe, domestic radio telephone service is becoming more and more common. Deluxe trains offer the traveler a hand set phone by means of which to carry on conversations wherever he will, even as the wheels carry him across the continent; and there is no practical engineering reason why American trains should not be allowed this same sort of service except that (1) it would involve use of both RCA and Bell patents, and (2) neither company will license the other to use its patents for such service. One never knows out of what minor program a major competitor may grow. That is just a minor bargaining consideration worked into the treaties, but there is no telling what may happen in the next year or so to destroy all these cozy arrangements.

Television is pressing for disposal. Under the agreement, television as an adjunct of telephony (so that conversationalists may see each other as they now do in Germany) is definitely assigned to the Bell system. But broadcast television, the amusement, the public service in its own right, the great new frontier in industry, is assigned specifically to neither in the treaty.

Whose shall it be? How will government allow the treaty of 1932 to affect the general public interest, necessity, and convenience? We know RCA is experimenting handily; and so are the Bell laboratories, with Philo Farnsworth as an independent ally. RCA is working to perfect the principle of transmission by way of the electro-magnetic spectrum, free of the Bell system's wires. The Bell system, with its new coaxial cable, is determined not to be excluded from new business for the wire network it has built and protected at such expense. The adjudication of patent rights and the determinations of the Communications Commission between applicants for licenses of operation, once standards of performance are set, will answer the question.

18. Patents and Power

A FRESHMAN IN AN IDAHO HIGH SCHOOL SHOCKS HIS CHEMIStry instructor by sketching out on the blackboard a complete conception of how to see by electricity. Two Russian émigrés, huddled over glasses of tea in a Second Avenue café, wonder where in all New York they can turn to commercialize a project of the same general type. The gods must have laughed that day when they set the impecunious ex-soldier of the Czar and a child in knee pants at each other. The gods started something which affects a great many more people than just those two, however. For Philo Farnsworth and Vladimir Zworykin are the symbols of power predicated upon invention, of fortune waiting upon the word of government. As between these two eventually must be decided basic rights under letters patent from the United States Government, rights of vital importance to the exploiter and user of television.

The axis of control upon which both the American Telephone and Telegraph Company and the Radio Corporation of America have developed is the patent. In the one case, all independent competitors were required to merge themselves into a single organization in return for licenses to use the original Bell patents. In the other, manufacturers of radio equipment found themselves unable to proceed without licenses from RCA, with the result that today more than ninety-five per cent of all the receiving sets in existence carry with them the extra burden of royalty payments to the licenser. Those who hope to control or share in the profits of television are hopeful that they can emulate such success. Nearly every Sunday newspaper has feature stories about this or that revolutionary discovery, just patented. But the decision as to whether these beliefs are well founded comes only after years of litigation and expenditures of considerable sums.

The patent itself is nothing more than a limited legal monopoly upon the use of a particular creation. The basis for grants of patents varies in detail among nations, but the general principle is to secure for the inventor a just reward for his ingenuity. In the United States, the tests for award involve priority of conception, novelty of thought, and utility. Once a patent is awarded, the holder is permitted a monopoly within the meaning of the particular grant. The monopoly is perfectly constitutional, but that does not resolve the strains and stresses that continue to center about its exercise. Monopoly has always been repugnant to everybody affected by it except the monopoly holder. As long ago as the fifth century A.D., the Byzantine Emperor Zeno decreed that no one might presume to exercise monopoly of any kind of clothing or fish or any other thing serving for food or any other use. He also forbade that any persons might combine or agree in unlawful meetings to fix the minimum prices for sales of merchandise. Zeno needed and sought popularity with the masses.

The struggle against monopoly was recorded in England as early as 1350, but it never was resolved finally one way or the other. Queen Elizabeth let a great number of crown charters for trade monopolies, and England flourished. But by 1623 the people were so indignant at the administration of these legal permissions that Parliament and James I were prompt in the declaration of the Statute of Monopolies which was intended to repeal many and lighten the effect of other crown charters. But monopoly, the ugly devil, would not be downed. He crossed the sea with the colonists of the New World. As late as 1933, the Government of the United States was still wrestling with him, and an experiment noble in purpose was made, by way of the National Industrial Recovery Act, to exercise his virtues and exclude his evils.

But never, throughout the long and complicated struggle to divine between the good and evil of monopoly, between stimulation of industry and protection of the consumer, has government shown any serious inclination to preclude an inventor from receiving reward for his novelty and priority of thought, always providing the thing of his conception can be made useful.

When the United States Constitution was being drafted, Benjamin Franklin, who fixed upon us the habit of calling electricity positive or negative, caused the insertion of a provision that "Congress shall have the power . . . to promote the progress of science and the useful arts, by securing for limited times to authors and inventors an exclusive right to their respective writings and discoveries."

James Madison, in *The Federalist*, commented that the utility of this clause in Article I, Section 8, could scarcely be questioned, as the copyright of authors had already been adjudged in Great Britain to be a right at common law. The right to useful inventions, he concluded, seemed with equal reason to belong to the inventors, and the public good coincides in both cases with the claims of individuals.

It would be interesting to have the comments of Madison and of Franklin upon the situation in which the inventor finds himself today, and the uses to which the patent laws are put. In their time, the inventor was still working within a handicraft economy. It was not impossible for him to fabricate with his own hands the conception of his mind and then peddle the product where he would. Nobody foresaw the day when large corporations would establish laboratories and pay inventors fixed salaries in return for an assignment to the corporation in advance of any patents attainable as a result of the endowed research.

Out of this system has developed the popular phrase "captive inventor," implying that he who thinks and tinkers for a corporation thinks unhappily, and tinkers only because he must eat. There is no evidence that such necessarily must be the case. It would appear that the inventor invents or he doesn't, and that the state of his finances has relatively little to do with the state of his intelligence. True, the charge that great laboratories are sterile of original thought is supportable to some extent. The fiery-eyed zealot who starves himself and pawns his wife's wedding ring while working furiously in a garret, appears to bring out more novel ideas than the well-fed researcher in the corporate laboratory. But in the machine age it is the laboratory and the corporation which must develop the original conception for practical use. A man may conceive a revolutionary principle for locomotion, but unless he is adequately equipped with plants and capital he cannot get his instrument into production.

And that is the dilemma in which our television experimenters find themselves, for their product, once priority of claim is finally adjudged, must always go in the end to some impersonal power, some corporation, for the beneficent, if profit-making, effects.

On that account we must have a background of knowledge concerning how a patent is finally determined. We must know how two claimants of a single conception settle their problem. First, of course, they apply at the Bureau of Patents for a recording of their claims. A popular misconception exists to the effect that once the government agency has issued letters patent, the exclusive rights are established forthwith and all one has to do is look up a financial supporter before launching into full-scale production. Actually, nothing of the sort happens. If the patented device is of any material importance, the verdict of ownership is decided ultimately in the Supreme Court of the United States. It is on that account that years and dollars are consumed, generally in direct proportion to the value of the patent.

The first patent law passed by Congress by virtue of its constitutional privilege was signed by President Washington on April 10, 1790. It was a simple law providing that a device that concerned "any useful art, manufacture, engine, machine, or device, or any improvement therein not before known or used" might be patented. The right was to last for 14 years. Administration resided in the Department of State, at that time headed by Thomas Jefferson. A patent board, consisting of the Secretary of State, the Secretary of War, and the Attorney General, was authorized to settle disputes. Jefferson, himself an inventor, was favorably disposed toward the law and held that a man ought to be allowed a right to the benefit of his invention for some time. "Nobody wishes more than I that ingenuity should receive liberal encouragement."

Until 1836 no serious changes occurred. At that time an act was passed to remedy the difficulties that the increasing number of patents was causing. The Patent Office was set up as a separate bureau to care for a systematic examination of inventions and determine to a limited degree the patent requirements of utility and novelty. In 1870, the patent law consolidated the previous acts, twenty-five in number, which had developed since 1836. The act of 1870 has become the basis of the present patent system.

Developments of the patent law have been largely procedural, neglecting to a large measure the country's change from a handicraft to a machine technique. Where formerly single patents covered the operation of particular devices, many are necessary today. An automobile, for example, is the mechanical result of combining many patent principles. If the inventor desires to exploit his invention he runs the risk of conflicting and interfering with others' claims. As a result there is a reasonable timidity on the part of the financier to undertake support of a patent not definitely insured against danger of conflict. After he invests in a factory, machinery, and the other things essential to operation, someone with a prior right of invention may be able to wreck the infringer's operation, hold him liable for damages, put him out of business, and even invade the homes of innocent purchasers with full legal authority to destroy every copy of the infringed patent.

The following story is typical of the threats surrounding the entrepreneur: "We went into our factory and if we tried to wind the coil this way somebody out in Oklahoma had a patent for it. If we tried to wind it another way somebody else in Peoria, Ill., had a patent for it; and if we decided not to wind it at all, we found omitting it was covered by a patent of somebody else." [1] Since there are more than two million patents extant in the United States it is obviously impossible to make an absolute determination, simply from Patent Office records, that there is no chance of infringement. Specific plaintiffs and respondents, necessarily, resort to the courts in separate instances, hopeful of settling general claims of priority and novelty.

The functioning of the system is considerably complicated by the large number of useless and absurd patents that have been granted. For instance, there is the pedal calorificator which is a device consisting of nothing more than "a set of tubes running from your nostrils to your feet for the purpose of keeping your feet warm with your own hot air in the winter." Another is an automatic derby tipper which saves the individual the trouble of lifting his hat when meeting a young lady on the street. The entire contraption weighs at least 15 lbs., and the major part of it resides inside the hat.[2] Sometimes, to escape this sort of thing, the Patent Office will demand workability from an operating model. This has prevented, so far, the patenting of perpetual motion devices; and for years it restrained those seeking monopoly upon flight. In fact, the Patent Office included both perpetual motion and flying in the same category for many years, and it was not until 1896 when Professor Langley flew his quarter-size, steam-driven model that the Patent Office considered flying a possibility.

The cost of litigation is, for practical purposes, prohibitive to the independent inventor and the relatively small manufacturer. It is estimated that the cost of continuing a patent suit through the Supreme Court begins at one hundred thousand dollars; and it was testified that a million was spent in protecting the Edison incandescent lamp against infringement, and that Edison spent more money in litigation than he made in royalties.[3]

These patent suits are among the most involved of all legal proceedings. Extensive search for the facts must be made; experts and attorneys must be hired; court records must be printed. Patents, therefore, may be said to have meaning only when supported by sufficient financial resources; and hence they are called by the cynical only the right to sue or be sued.

The patent system has become, for all practical purposes, the playground of the large corporations which establish a legal position to frustrate competition, develop a degree of protection against technological change, and sharpen a weapon for trading with other corporations. It has been charged, but without any conclusive proof, that these corporations buy up patents wholesale and, in pursuance of a trading philosophy, deliberately withhold inventions from the public. A more accurate charge, probably, would be that corporations delay the output of a newly patented device, or suppress productive progress for a time, while trying to organize their economic status to advantage. This is but a natural concomitant to their very existence. The purpose of a corporation is to make money. The purpose which motivates an inventor is not so clearly defined. He has the instinct for contrivance. He exercises it. Whether he is spurred by desire for money is a question not possible of categorical answer. Some inventors have exhibited intense interest in money, others show no interest at all in money as such. But whatever the intent of the inventor or the corporation, the intent of the law has been as much to protect the public welfare as the inventor or the corporation. The Supreme Court, in one of its very earliest decisions, stated that while one great object of the constitutional provision was to hold out reasonable rewards to inventors by guaranteeing them exclusive rights, the ultimate purpose was to promote the progress of science and the useful arts.[4] This would seem to forbid conscious withholding of patent usages.

However, not even the Supreme Court has been able to maintain a consistent view of what the limitations of the inventor's rights may be. As the individual's powers have been absorbed by corporations and the exercise of patent rights has become a standard business practice in corporate management, with profit as the initial motive and promotion of the arts and sciences at least apparently secondary, the Court has shifted its stand.

A most important redefinition of rights was made in the case of the suits involving the original Bell patents and attempts by the Federal Government to prevent continuance of the Bell corporate licensing program. The Court pointed out that each invention has separate rights, even though held by a single inventor. The invention loses none of these rights even though successful operation depends upon its being used in conjunction with other devices, which may or may not be protected by patent.

All that the patent law requires is that when a patent expires the invention covered by the patent shall be free to everyone, and not that the public has the right to use of any other invention, the patent for which has not expired, and which adds to the utility and advantage of the instrument made as the result of the combined inventions.

Counsel seem to argue that one who has made an invention and thereupon applies for a patent therefor, occupies, as it were, the position of a quasi-trustee for the public; that he is under a sort of moral obligation to see that the public acquires the right to the free use of the invention as soon as is conveniently possible.

We dissent entirely from the thought thus urged.

The inventor is the one who has discovered something of value. It is his absolute property. He may withhold the knowledge from the public and he may insist upon all the advantages and benefits which the statute promises him who discloses to the public his invention.[5]

In other words, the constitutional directive no longer is primarily to advance the useful arts and sciences but to protect the trading position of the individual or the corporation. It is with this conception of the law that we must approach the specific patent problems of television.

In no industry are they more complex. Literally thousands of patents are exercised in combinations and exchange agreements, cross-licenses and by simple consent, to effect a single program. The principal patents, known in the jargon of the corporate law as "controlling," the ones without which no part of operations can proceed, are particularly difficult to determine. The difficulty was brewed back when that schoolboy in knee pants and that Russian émigré in a New York café began to scribble their conceptions of television upon blackboard and tablecloth. The schoolboy today is known as the famous inventor, Philo Farnsworth. Vladimir Zworykin is the principal "captive" of the Radio Corporation of America. Farnsworth is a participant in British television, having entered his patent claim in an open patent pool in England as required by the government there. He has an agreement with the Fernsee organization in Germany. In the United States he has licensing agreements with Philco, the American Telephone and Telegraph Corporation, and Columbia Broadcasting System. But Philco is a licensee of the Radio Corporation of America, bound to turn over to RCA any novelties created within the remotest extension of that license. Zworykin, though now one of RCA's principal inventors, once worked for Westinghouse Electric and Manufacturing Company, which today holds his original claims as basis for suits not only against Farnsworth but RCA as well. Add to this involved situation the claims of the American Telephone and Telegraph Company, of the thousands of individual inventors around the country with sufficient funds to demand a court test, and you may get some idea of the task ahead of the Supreme Court.

These may be fascinating prospects of pyrotechnics for the public and profits for the counselors, but within the radio industry the patent situation is literally a matter of life and death. Samuel E. Darby, Jr., attorney for the "independent" radio manufacturers, has already unleashed a barrage against RCA.

The Radio Corporation of America, by reason of the pooling of relevant patents virtually of the entire electric industry, is in control of broadcast transmission and the manufacture of radio receivers, and one question to be considered is how far that control will be allowed to be extended into the television field. . . .

In other words, anyone who wishes to engage in the radio business today or in the television business tomorrow, must ask and get the permission of RCA.[6]

Philo Farnsworth, it may please Mr. Darby to learn, has won the first skirmish. On July 22, 1935, the Patent Office awarded him priority in an interference action with Zworykin, and he was sustained on appeal in the same office on March 3, 1936. But a long road is yet to be traveled before the Supreme Court makes the ultimate decision.

Farnsworth, a young man as inventors go, has set himself up as the great independent in television research, but actually he is no more independent than the dollars of his financial backers. With their help he is able to contemplate with a fair amount of equanimity the difficulties of his opponent, Zworykin, who on July 9, 1936, again found himself in patent litigation, this time in a contest between RCA and Westinghouse concerning the patents of Zworykin and Henry Joseph Round, of RCA Laboratories. Zworykin, during the years 1923–1925, made certain developments in television research which he assigned to Westinghouse, then his employer. These were in conflict with developments by Round over which RCA has control. (By way of explanation it should be stated here that all Zworykin's recent developments belong to the Radio Corporation only. The inventions in question in the case mentioned concern his developments while he was with Westinghouse.) Previously the Patent Office had decided in favor of Round's developments, so Westinghouse took the case into the law courts.[7]

The contest between RCA, Westinghouse, and Farnsworth obviously will have an important part in determining the direction and control of television. However, this case is important not only because it is one of the major focal points of conflict, but for the insight it furnishes into the workings of our patent system. The impecunious inventor, the history of this affair clearly shows, must depend upon the wealth of large corporations and the refinements of means in the laboratories of these corporations to carry a conception through to successful commercial ends. In the Farnsworth-Zworykin case before the patent examiners, the testimony of both men in defending their rights to the priority of their respective patents is astonishing:

Q. To what does the invention in issue broadly relate?
Farnsworth: To a transmitter tube for television.
Q. When did you conceive the broad idea of the subject-matter here in issue?
Farnsworth: About March 1922.
Q. Where?
Farnsworth: At Rigby, Idaho.
Q. How do you fix this date?
Farnsworth: At that time I was a Freshman in High School. I fix the date largely by the fact that at that time I was being permitted to take a course in chemistry, which was not usually followed in the Freshman year. In fact I . . . started in mid-term. That required that I make up the work for the extra term, so that the time lies some time between December, 1921, and the time when I left school, which was May 1922. . . .
Q. Did you discuss it [the television idea] with anybody?
Farnsworth: I discussed it with Mr. Tolman, who was tutoring me in chemistry.
Q. About what time did you have this discussion with Mr. Tolman?
Farnsworth: There were many such discussions during the period that he was tutoring me. I place the first one as

near as I can about the middle of that period from December to May, approximately March. . . .

Q. When did you leave school?

Farnsworth: Early in May, 1922. I left school to help with the early farm work, a little bit before the regular closing time of the school. [Farnsworth's formal education never did carry him further than intermittent attendance at Brigham Young University and reading in the library there.]

Q. Did you make at that time a written description of the invention?

Farnsworth: It was the practice of Mr. Tolman and I to make sketches and diagrams mostly on the blackboard, but at times also on a scratch pad.[8]

Farnsworth went on to say that he left his parents' farm in 1922 to take a job as electrical helper at Glenn's Ferry, Idaho, at fifty cents an hour. His earnings were used "supporting myself mostly" and saving for a time when he could go back to school. He then went to Provo, Utah, to work in a machine shop and foundry. "I attempted at that time to patent and to obtain money to promote a rectifier for use in radio programs, this all with a view to obtaining in some manner or other means of developing this television idea."

But the venture proved a failure and Farnsworth lost the one hundred and fifty dollars provided by his father; a tremendous setback for both son and parent.

In 1926, after five years in which the television idea was suppressed through the circumstances of poverty, he found financial support. Two men by the name of Gorrell and Everson put up some money and formed a partnership with Farnsworth to develop a laboratory in Hollywood, California.[9] It was in this year that Farnsworth applied for a patent. And from then on Farnsworth has become a potent figure not only in the technology but also in the business of television. After eight years of work on the device, Farnsworth said, he found no essential differences between the patented invention and the conception originally presented by the fifteen-year-old high school freshman to his instructor.

After success in his patent-interference action against Zworykin in the Patent Office, Farnsworth received support from the powerful corporate systems of A. T. & T., Philco, and Columbia. Should they decide to finance him to the finish, both as to technical development and legal patent protection, Farnsworth will be in a position to overcome the specter of infringement suits by RCA or any other competitor.

The story of Zworykin's struggle is hardly less exciting than Farnsworth's. A radio expert for the Signal Corps in the Russian Army during the World War, he worked with Professor Langwin on x-rays and electrical and gaseous discharges, and with the Russian Society of Wireless Telephone and Telegraph. When Russia was turned upside down by the Revolution, Zworykin drifted to the United States, the land of promise, in 1919. He had conceived his idea back in Russia in 1917, and when he arrived in the United States he discussed it with a friend named Mouromtseff. "In fact, I even proposed to Mr. Mouromtseff to organize a development of television according to my system in America, but we both didn't have any money to start this and therefore the project did not materialize."

Zworykin, like Farnsworth, found making a living a difficult preoccupation. "I was looking for a job but couldn't find any and departed from New York to Omsk, Siberia, about the middle of March, 1919."

Q. When you were in New York were you employed, that is, I mean did you procure any employment after your return from Siberia?

Zworykin: Not in the first couple of months after my return. I tried to first find somebody who may be interested in my invention, but failing that, I obtained a position as bookkeeper with the Financial Agent of the Russian Embassy.... Probably in October, 1919.... About one year.

Q. Why did you choose employment as a bookkeeper?

Zworykin: That was the best I could obtain, and besides, Mr. Mouromtseff helped me secure the position.

At the end of the year with the Russian Embassy, Zworykin received the position of research engineer with the Westinghouse Electric and Manufacturing Company. After another year, he left to go with the C. & C. Developing Company in Kansas City. In 1923, he returned to the Westinghouse Company and later, some time in 1928, he became associated with the Radio Corporation of America.[10]

Westinghouse owns some of Zworykin's creations, and RCA others. In 1931, the Westinghouse Company brought an interference action in the Patent Office concerning certain television patents developed by Zworykin which Westinghouse owned and claimed were prior to the inventions of Henry Joseph Round, of the RCA.[11] The basis of this suit is not easy to understand since Westinghouse and the Radio Corporation are parties to a broad cross-licensing agreement in which each has reciprocal rights to the other's patents. However, Zworykin was in the peculiar position of defending his own creations, owned not by himself in any particular but by Westinghouse, against his present employer, RCA. The result of this action was a decision by the Board of Appeals of the Patent Office on January 10, 1936, granting Round priority. On July 9, 1936, Westinghouse filed a patent suit in the U. S. District Court of Delaware against the Radio Corporation of America on this issue. Who shall win? Only time and the Supreme Court can decide.

But, however the issue of basic patents is decided, the fight is not settled. There remains the question of organization for operation. Standards of performance must be fixed which involve one set of patents and methods as against another. This, we know, is a matter which must be disposed of by the Federal Communications Commission.

19. Past Is Prologue

WHAT IS THE VALUE IN REVIEWING THE PAST?

De Forest has faded from competition with RCA, which bought up his bankrupt plant, and most of his colleagues have gone the same way. The Bell system has settled its differences with RCA in sound motion pictures and neither offers to compete with the other in broadcasting or in domestic telephony, but television is another matter. The amended treaty of 1926 fails to dispose of it in a clearcut manner, and we know that the acrimonious exchanges of Mr. Jewett and Mr. Sarnoff concerning the respective values of the coaxial cable and the radio spectrum indicate that neither intends to allow the other to dominate.

Where stand the remnants of competition? What may be expected of that people's champion, the Federal Communications Commission? Is the struggle for television to be another exhausting battle such as that which we have recounted in sound radio and sound motion pictures? If the radio were not so intimate a force with the American people, or if the American people were more intimate with the forces that have controlled the development of radio so far, it might be unnecessary to have any concern for these matters.

Of three business institutions with most at stake, we find one somnolent even though warned by a Paul Revere who certainly could not be said to have detoured headquarters. Such is the case of the motion picture industry, which shows no apparent interest in the report on television given it by A. Mortimer Prall, whose father was the chairman of the Communications Commission. There is nothing somnolent about the Bell system. It is divesting itself of the sound motion business by selling ERPI and settling numerous anti-trust suits by independents in that industry. Like a champion boxer, it is poised for action. A cable is already laid between Philadelphia and New York, and rules of service for television and telephony have been made, one of the most interesting of which is the Communications Commission requirement that wires be used instead of wireless for relay of programs wherever that is at all possible.[1] And there is nothing somnolent about the Radio Corporation of America. If anything, its conduct is feverish. Unlike the Bell system, it has failed to soothe those whom it has been unable to destroy, and it has failed to destroy some who thirst for its blood.

There is, for example, the Philco Radio and Television Corporation, generated by the Philadelphia Storage Battery Company as a corporate life-saver in 1927 when radio was converted to use on ordinary 110-volt house current. Philco, we know, is a television licensee of Philo Farnsworth (the "young De Forest"). But Philco is also a licensee of RCA in radio. It is an extremely vigorous licensee, too. Nearly destroyed when general need for radio batteries was ended by technological advance, Philco has come back so strongly that it has sold more receiving sets than RCA in equal periods of time. In the good year of 1934, for example, Philco sold 1,250,000 out of 3,550,000 sets bought by the American public. RCA, which had given Philco its literal lease on life, sold a mere half million as runner up.[2]

By 1936 RCA was wondering, quite naturally, how on earth to restrain this galloping infant competitor it had loosed against itself. True, on every Philco set RCA received a royalty, but nothing relieved the strain upon RCA's own investment in manufacturing plant which was being assaulted bodily by loss of sales, or upon corporate pride. The obvious thing to do was to terminate the Philco license. But should that be done without some certainty of just what Philco was doing in television?

On July 30, 1936, Philco brought suit against RCA, the RCA Manufacturing Company, John S. Harley, Inc., Charles A. Hahne (or Hahn), and Laurence Kestler, Jr., charging them with unfair, wrongful, and illegal methods and practices, including the use of subterfuge, deception, false representations, and efforts to corrupt Philco employees and employees of Philadelphia Storage Battery Company by inciting them to breaches of trust and confidence reposed in them, in an "endeavor to entice, bribe, persuade and induce said employees to divulge and procure for them confidential information, data, designs and documents. . . ."

Hahne and Laurence Kestler, Jr., were accused of entering Philco's factory and therein and elsewhere putting themselves on good terms with numerous girls and young women in the employ of Philadelphia Storage Battery Company. This is in the spy tradition, but, reversing the tradition of spies among nations in which beautiful girls wheedle secrets from handsome young soldiers, the radio men took the

Philco girls over to see the bright lights of Philadelphia and then, according to the language of the complaint:

> Did provide them from time to time with expensive and lavish entertainment at hotels, restaurants and night clubs ... did provide them with intoxicating liquors, did seek to involve them in compromising situations, and thereupon and thereby did endeavor to entice, to bribe, persuade and induce said employees to furnish them for use by all the defendants, confidential information and confidential designs, all in breach of the duty of trust and confidence which said employees owed to the plaintiff herein and to said Philadelphia Storage Battery Company.[3]

This is only one of the more humanized passages of the Philco complaint asking the Supreme Court of the State of New York for relief from such actions. The others deal with more complex legalized aspects. And in Wilmington, scene of the old fight concerning De Forest and the cross-licensing agreements, RCA had to meet Philco on a second action seeking to restrain it from withdrawing the Philco license. RCA's publicity agent had just crowed, "our patents, which include the iconoscope and kinescope, have secured for the United States world supremacy in television." But in answering Philco's suit RCA's lawyers stated that, in seeking to extricate itself from agreements with Philco, RCA was only trying to forestall a nasty television patent problem. This would appear to be a contradictory state of affairs.

Philco is not to be exorcised by responses to lawsuits. It has just acquired license for a new television transmitter to operate with 15 watts of power in the 204-210 megacycle zone, and it still demands secure tenure of licenses and inviolate relationships with its employees. It still defends its trade secrets fiercely, and it continues to turn out more and more sound radio sets. And it keeps Philo Farnsworth snuggling closely, for all that it may have understandings with RCA and the Bell system. Philco must not be dismissed from the mind.

But whatever happens to Philco, there is still to be considered the matter of the Columbia Broadcasting System. The CBS is a program organization, pure and simple. Though it owns a few station licenses, in the matter of trade practices, it has an extremely high rating within the radio industry. With the public it is also relatively high in favor because of its "sustaining" features offered to fill in program time-space not sold to some commercial sponsor. The most famous of these has been for years the Sunday afternoon broadcast of concerts by the New York Philharmonic Symphony Society against which Mr. Sarnoff has only just lately countered with his NBC Symphony conducted by Arturo Toscanini.

Columbia is now installing a sound-sight transmitter in the Chrysler Building, to operate on a frequency in excess of 40 megacyles, with peak power output of 30 kilowatts. Its radius of reception will be about forty miles, and the definition exceedingly clear—about sixty frames of four hundred and forty-one lines each per second. It has retained Gilbert Seldes as program director, with a view to making television programs as good as those sound radio offers today. William S. Paley has announced that two million dollars would be spent to develop Columbia's technique of television broadcasting.

The trade magazine *Business Week*, wishing Columbia

well, pointed out that thanks to its control of basic patents, the Radio Corporation of America collects a license fee on every radio set manufactured in the United States. For, it pointed out, RCA could legally force the stoppage of the whole thriving set manufacturing business, if it wanted to, by refusing to renew licenses as they terminate.

The set makers entertain golden dreams of tomorrow's harvest when television becomes a commercial reality. But RCA is out to win the same dominant position in television that it holds in radio; and that disturbs the hopeful dealers.

The set manufacturers together with the broadcasting companies that entertain a similar concern about radio and television sending equipment, argue that a little competition might ease the situation; even two masters would be better than one.

It is because of these sentiments that the trade was so pleased last week with the Columbia Broadcasting System's announcement of its plans to install a powerful television transmitter atop the Chrysler tower.[4]

Evidently the enthusiastic seekers after competition were more eager than discerning. They should have inquired who manufactures Columbia's television equipment. And they should have inquired who is going to transmit the programs from station to station, for Columbia has no independence there. In the one case it must turn to RCA for radio relay equipment. In the other, the Bell system furnishes cables. And how do Columbia and RCA stand with the Bell system in the matter of using that transmission equipment?

The treaty of 1932 provides that RCA may use the Bell facilities for wire program transmission, picture transmission of material for programs, electrical sound recordings, one-way transmission of current for control of frequencies, and systems for radio program transmission or wire program transmission. On the other hand, the contract with the Columbia System provides that the facilities furnished by the telephone company are only for use in one-way radio program transmission. . . .[5] And so there goes our competitor, tangled in clauses and whereases worse than ever was Laocoön with the serpents.

What, then, has RCA to fear? It has fended off anti-trust suits and patent suits. It has its greatest competitor in set manufacture (Philco) on tenterhooks. It has Columbia, its great competitor in programs, buying RCA equipment and adversely placed in relation to RCA on the Bell transmission system. Mutual Broadcasting, the third largest program service in sound radio, has developed no known position of importance in television.

So, again, what has RCA to fear? There is always danger that someone in authority may hold that it is not serving the public interest, necessity, or convenience. And there is evidence of restiveness toward RCA. Representative W. D. McFarlane, of Texas, rose before Congress on July 19, 1937, and attacked the monopoly characteristics of the radio industry in particular; and on August 10 he made a second speech, going into detail concerning both Columbia and RCA.

It was exactly in this same way that the "radio trust" was attacked in 1929. When Mr. McFarlane spoke, the great Roosevelt boom was at its richest flower, just as the Hoover boom had been in 1929, and the radio industry, as before, only smiled as he demanded inquiry into its activities. But before 1937 was ended another business depression had set in as one had set in toward the end of 1929, and the radio operators were becoming alarmed, if belatedly so, for they are sufficiently skilled in public psychology to know that depressions spur Congresses to "investigate." The statesmen hope, somehow, by taking testimony and making findings, to exorcise business miseries. And resolutions to investigate radio were before House and Senate.

Mr. McFarlane's speech of August 10, 1937, "Radio Monopoly Must Be Curbed," may be the unnoticed turning point in a new national policy concerning electronic communications, or it may lead to nothing. We quote the essentials of it here. They should be considered against the background of facts the reader already knows as significant indicators of what passes through the mind of the nontechnical radio critic in public office:

An analysis of the board of directors of the Radio Corporation of America bears witness to the correctness of the remarks of my colleague from Texas, Mr. Patman.

Gen. James G. Harbord is a Morgan representative on the board of the Radio Corporation of America and is also a director of the Morgan-controlled Bankers Trust Co. Newton D. Baker is legal adviser to many of the Morgan-controlled utility companies. Cornelius Bliss is a member of the firm of Bliss Fabyan Co., a Wall Street firm, and is also a director of the Morgan Bankers Trust Co. The elder Bliss was for many years treasurer of the Republican National Committee. Arther E. Braun, of Pittsburgh, is president of the Mellon Farmers Depositors National Bank, one of whose directors is A. M. Robertson, chairman of the Westinghouse Co. . . .

Bertram Cutler is listed in Poor's Register of Directors as

being connected with John D. Rockefeller interests. Edwin Harden, the brother-in-law of Frank Vanderlip, is a member of Weeks & Hardin, a Wall Street firm. Dewitt Millhauser is a partner in Speyer & Co., underwriters of utility issues. Frederick Strauss represents J. W. Seligman & Co., a Wall Street firm. James R. Sheffield is a corporation lawyer, a former president of the Union League Club and the National Republican Club. As a former Ambassador to Mexico he used his political connections with the Hoover-Coolidge State Department to get concessions for R.C.A. in South America.

Although the control of the Columbia Broadcasting System is supposedly a Paley family affair, the bankers are not without influence. When the Columbia network was purchased back from the Paramount Picture Co., the representatives of the financiers who put up the money for this purchase were added to the board. In return for the cash which the bankers put up they received approximately 50 percent of the Columbia Broadcasting System's class A stock. These banking interests were Brown Bros., Harriman & Co., W. E. Hutton & Co., and Lehman Bros. The members of the board of directors who represent these bankers are Prescott S. Bush, partner in Brown Bros.; Joseph A. M. Iglehart, partner in Hutton & Co.; and Dorsey Richardson, of Lehman Bros.

At this point I should like to say something about the Radio Trust formed by R.C.A., General Electric, Westinghouse, A.T.&T., et al., and which was supposedly dissolved by the Government in the notorious consent decree of 1932. Before the consent decree, R.C.A., who, under the illegal cross-licensing agreement with A.T.&T., et al., controlled the patents to radio-equipment manufacture, began to issue licenses to others—probably with the idea of convincing the Government and the public that they were not such a bad trust. But, after the consent decree, I have learned of no licenses for radio-set manufacture that were given by R.C.A.

When the Government seemed to be pressing suit against the Radio Trust, the cost of radio sets dropped to $10 and below. This permitted millions of homes to enjoy the benefits of radio, and millions of people were able to listen to the issues of the day aired over the wave lengths. A new note in democracy was being struck. However, just as soon as the Hoover administration and the Radio Trust entered into the now infamous consent decree the price of radios began to rise again until now $30 and up is the price for a decent radio.

Not content with their monopolistic control and the 5 percent on gross revenue they take from all licensees they began to terrorize even those who had licenses to compete with them. The case of Philco Radio & Television Co., which filed a suit against the R.C.A., charging espionage and other terroristic practices to R.C.A. is eloquent testimony.

Other independents, if they desired to compete, were forced to run the gamut of patent-infringement suits brought by R.C.A. To fight a case of this sort costs a great deal of money. The adjudication of a patent through the Supreme Court sometimes costs over $100,000. Such a cost is prohibitive to most independents. His choice is due in one of two directions: Either he fights and the cost of litigation plus threats to his customers drives him out of business; or, he wisely goes out of business upon the receipt of a threat of an infringement suit. In either case, the independent gives up the ghost. Such is the power of the patent racketeering of the Radio Trust. . . .

In the supposed dissolution of the Radio Trust by the consent decree in 1932, it was proven that R.C.A. possessed such a monopoly. There is evidence to show that despite the consent decree, this monopoly still persists in violation of the anti-trust laws. Yet testimony before the House Appropriations Committee shows that broadcasting licenses of R.C.A. are renewed every 6 months without ever having the question of the apparatus monopoly or public interest raised. I sincerely believe that the issue of reëxamining the effects of the consent decree is resting squarely on the shoulders of Congress. Shall we face the issue or evade it as has been the custom in the past?

The gentleman from Massachusetts [Mr. Wigglesworth], who, as a member of the Appropriations Committee, has given much time and consideration to this subject, has spoken several times favoring the immediate clearing up of this communications monopoly. His work in the committee bringing out the existing known facts, I am sure, has the hearty approval of the Congress. Several other Members have spoken, pointing out the great need of an investigation. . . .

Mr. Voorhis. Mr. Speaker, will the gentleman yield?

Mr. McFarlane. Yes; I yield.

Mr. Voorhis. Does not the gentleman feel that perhaps the root of this whole matter is to be found in the fact that these corporations have been able to call a certain radio channel their own; that, as a matter of fact, if there is any natural resource that ought to belong to the people it is the air, and that we are gradually building up here a vested interest in the ownership of channels of communication through the years? Would not the gentleman favor some tax measure which would levy a good stiff franchise tax and take the water out of the situation so that the only advantage would be a temporary license, or a license running for a certain period of time? Would not this prevent the building up of a vested interest in these channels?

Mr. McFarlane. Answering the gentleman, I may say that there has been tax legislation pending before the Ways and Means Committee since the early part of this year, but we have been unable to get any action on it. This would require the radio industry, which is the only public utility operating in interstate commerce in the United States today that does not at least pay the cost of its supervision, to pay a suitable tax; but this bill, like the others which should have been brought to the floor of the House, never has been considered by this committee and still lies buried there.

Mr. Leavy. Mr. Speaker, will the gentleman yield?

Mr. McFarlane. Yes; I yield.

Mr. Leavy. The gentleman's remarks indicate that he has given much thought and study to this question, and he is making a strong case. I am wondering if he has covered the further abuse that is generally recognized of large, metropolitan newspapers of the country acquiring radio stations and then hooking in with the great radio chains and thus controlling channels of news through radio as well as through the press?

Mr. McFarlane. If the gentleman will read my remarks of July 19, he will see that I dwelt upon that very question; that I pointed out that some 200 of the large daily newspapers of this country own the largest radio stations in America, and through this method of radio broadcasting and sound motion-picture equipment and through the press, through that tie-up, they absolutely control and mold public opinion in this country today; and this is why Congress is having such a terrific fight to get any worth-while legislation enacted for the benefit of the people. [Applause.]

Mr. Wearin. Mr. Speaker, will the gentleman yield?

Mr. McFarlane. Yes; I yield.

Mr. Wearin. That tendency on the part of the newspapers coupled with the operation of the present chain does constitute a serious threat in the way of a monopoly to influence public opinion, does it not?

Mr. McFarlane. There is no doubt about it.

Mr. Wearin. I am sure the gentleman is familiar with the fact that I have a bill now pending before the Committee on Interstate and Foreign Commerce to prevent a continuation of this monopoly.

Mr. McFarlane. I know the gentleman has had such a bill pending for some time, but he does not seem to be able to get action on that any more than the rest of us are on these other bills. We cannot, apparently, get these bills out of these committees which would be of such tremendous benefit to the people. And this communications monopoly is becoming more powerful all the time. Until now many Members dare not speak their sentiments against it, lest they be opposed by it for reëlection. . . .

But does the grasping of the monopoly stop there? Let me quote the following from the Hollywood Reporter of July 1937:

R.C.A. Now Believed Aiming to Control Communications

Washington.—There is a well-authenticated report that the Department of Justice is now willing to withdraw its objections to a merger of Western Union and Postal Telegraph. In inside circles this is seen as an indication that R.C.A. is moving to control the entire communications field.

The ultimate battle, of course, will come over the control of commercial television. In view of President Roosevelt's determination for a unified communications system, it is possible that if the big wire companies merge, R.C.A. might let the merged outfit have the communications business and devote itself to the amusement field and broader television activities. However, this possibility is not credited by those in the know.

They believe that R.C.A. will make every effort to control both Western Union and Postal in an effort to broaden its telegraph business, and that the fight will then be between R.C.A. and A.T.&T. for full control of both communications and television.

It is not thought possible that if the wire companies do merge, the new company would be able to protect itself against the threat of radio competition by acquiring R.C.A., the supremacy of R.C.A. being seen as much more logical. In any event, it is believed that the merger would make commercial television much more imminent.

There is an interesting sidelight to the relation between R.C.A., G.E., and Westinghouse, but nevertheless important, and bears mentioning here.

When General Electric, Westinghouse, and R.C.A. were busy dividing up the radio field amongst themselves a very peculiar transaction took place. In return for certain stock and physical assets given to R.C.A. and which R.C.A. itself valued at $42,864,812 plus the exclusive manufacturing rights and the royalties to radio device field, General Electric and Westinghouse received 6,580,375 shares of R.C.A. stock, the market value of which was $263,215,000. In other words, R.C.A. paid $220,350,147.50 for the exclusive rights in the radio-apparatus field, and gave the control of R.C.A. to G.E. and Westinghouse. The facts are borne out in an unchallenged affidavit on file in the Federal court. It is difficult to believe that they were worth that much. It is far easier to imagine the innocent investing public who owned R.C.A. stock, through no choice of their own, made a gift of these hundreds of millions of dollars to Westinghouse and General Electric. And from the message I read to you earlier from S.E.C. the law is unable to cope with this manifest racketeering.

I want to ask that the committee now investigating tax evasions and tax loopholes investigate this gift of $220,000,-000 to Westinghouse and General Electric and learn just what taxes were paid on this $220,000,000. I also ask that they report their findings to this body.

The people of the United States have paid $2,262,375 last year to regulate the communications industry. In all other kinds of industries operating under Government franchise the cost of their regulation is placed on the industry. Why, then, should the taxpayers continue to keep up the cost of the Federal Communications Commission? I think it is now time for Congress to shift this burden from the shoulders of the taxpayer on to the communications industry, which operates under Government franchise for which they pay nothing.

I cannot repeat too often the query, "What does Congress intend to do?"

It is a wise monopolist who knows when to ease up. RCA may not be the sort of institution that Mr. McFarlane and his colleagues think it is. All these things may have been said in misunderstanding of the facts, but there the facts stand and the opinions with them. The stockholders of RCA, battered as they are by the years of tribulation and lawsuits, have the record to ponder. The public, who may be called on to finance the development of television either through direct governmental subsidy or by purchase of equipment at original high prices, have some things to ponder, too.

They may think upon the reference in Mr. McFarlane's speech to the prices paid for receiving sets during and after governmental anti-trust actions, and upon the fact that RCA, like every other radio operator, must depend finally upon the "public interest, convenience, or necessity" for its license to exist in the broadcasting industry. RCA could continue to make equipment if barred from interest in stations, but it wouldn't be happy under such circumstances.

20. Return of a Pioneer

TELEVISION IS KNOWN WITHIN THE TRADE AS A "LOCK AND key" business. Transmission and reception are bound together in a mechanically monopolistic way, no matter what the courts or commissions say about legal monopoly or its absence; and there seems no way to extricate them from their relationship.

Of course, sound radio is a lock and key business, too, to a certain extent. Without an adequate receiver in operation, the broadcast is futile. But in sound radio, selectivity of programs is not very difficult within the framework of standards now developed. The average receiving set can tune in on from ten to a hundred broadcasting stations. Its dial spins with the world. But that cannot be the case with television as we know it according to present engineering development.

But Mr. Jewett, Mr. Espenschied, Mr. Sarnoff and Dr. Jolliffe have made it emphatically clear that the spectrum does not accommodate television broadcasts in either number or range comparable with sound radio. Today, the average city is served with three to seven sound broadcasting stations; but tomorrow these will be gone, and only one, two, or in the rarest of instances, three television programs will be available.

In the second place, the technical nature of television does not allow any variations in equipment comparable with those of sound broadcasting. We now have inexpensive little radios for bedroom tables, "high-fidelity" console types for the drawing room, and special kinds for automobiles. They vary in sound definition without losing entirely their ability to compete with one another in the actual reception of the radio signal.

Not so with television. If a program is scanned at the rate of four hundred and forty-one lines, sixty frames per second with RCA's iconoscope, then no set can translate the electronic impulses back into comprehensible pictures except one designed especially for reception of a four hundred and forty-one line, sixty frame iconoscopic broadcast. And so it goes. If scanning is done by use of pierced disks, helical arrangements of mirrors, or any other variation of mechanical systems, then only receivers geared to these scanners can function.

It is immediately apparent that some basic standards must be set: television must be all of one thing or another, technically speaking, if it is to arrive commercially. And so, when we consider all that has occurred in sound radio, we recognize the enormous responsibility placed upon the group which sets those standards. These technical qualities will, in the end, resolve all questions of television competition; and monopoly by exclusive patent holders will give them dominating positions. There are two trade associations in the radio industry which speak for all competitors in general in the resolving of these pressing questions, just as counsel and legislative friends speak for interests in particular. These two trade bodies are the National Association of Broadcasters and the Radio Manufacturers' Association.

NAB is the spokesman for the disseminators of programs. It encompasses more than sixty-five per cent of all station operators, and these do in excess of eighty per cent of all advertising business in radio. At the F.C.C.'s engineering conference in 1936 James W. Baldwin, managing director of NAB, suggested on behalf of his organization a plan of allocation which would provide eight television channels below 100 megacycles, but pointed out that they would not be enough, really, for the demand.

There are, however, more than technical considerations involved here. The American broadcasting system is a competitive system. It is a great system because it has been competitive. . . . [A relative term, you will recognize.]

And our plea is today that you allow television to develop on the same basis. Better we delay the introduction of television than in enthusiastic haste inaugurate it and find that through control of patents so powerful an instrument is in the hands of too few people.

If television were ready to be inaugurated on a basis of a national competitive service, he argued, then the F.C.C. was clearly under a very great responsibility in determining in advance whether for all practical purposes the ownership of basic patents and agreements, if any, between patentees would permit competition in the construction of television transmitters and receiving sets. He seems to have an unassailable position there.

We should also know in advance what relationship, if any, may be established between the sending and receiving apparatus. Will there be freedom in the selection of receiving sets or will the use of terminal facilities be controlled in a manner comparable with the telephone?

Surely everyone will agree that those who own television patents are entitled to a rich reward for their creative work, but because of the public service inherent in television, patentees should be denied the right to control its use. Keep it from the hands of monopoly and allow it to develop only on a national competitive basis.[1]

But just how valid, in view of the facts, is the chance of competition? Technically, television is a lock and key operation. Ownership of lock and key is decided on the basis of a patent position. A patent position, we will all admit, can be developed more easily by the rich and politically powerful than the poor and weak. And once a patent position is attained, the holder of a patent has the right, under the Constitution of the United States and the findings of the Supreme Court, to make or not to make the article patented, to lease or not to lease rights to others.

What Mr. Baldwin asks for, essentially, is an "open patent pool," of the sort ordered in Great Britain when television became a public institution in 1935. Such a group must allow anybody to join it who has a patent of value to contribute. And the contributor thereupon has common power with all the other participants to use any combination of the pooled patents he so desires to make a set of his own design, paying royalties to the particular contributors whose patents he happens to use.

If such a patent pool could be arranged in the United States, who would participate? Already, the sound radio industry is dominated by two basic organizations, the A. T. & T. and RCA. Could there be true competition so long as these two major operators continue to follow the lines of policy indicated in the treaty of 1926, as amended in 1932? Hardly.

But what about the others who attended the engineering hearing with Mr. Baldwin? Said the Radio Manufacturers' Association spokesman, James M. Skinner, who happened also to be the president of RCA's troublesome licensee, Philco:

> RMA has tried to crystallize the basic needs of television in a five point plan:
> 1. One single set of television standards for the United States so that all receivers can receive the signals of all transmitters within range.
> 2. A high definition picture approaching ultimately the definition obtainable in home movies.
> 3. A service giving as near nationwide coverage as possible.
> 4. A selection of programs, that is, simultaneous broadcasting of more than one television program in as many localities as possible.
> 5. The lowest possible receiver cost and the easiest possible tuning, both of which are best achieved by allocating for television as nearly a continuous band in the radio spectrum as possible.[2]

We are thoroughly familiar, by now, with Mr. Skinner's problems and objectives. One set of television standards—we know this is essential for uniform reception of the sort now common in sound radio. We also know it entails inevitable monopoly. A high definition picture—this is simply a test of consumer interest. Unless the picture is large, clear, easy on the eyes, television naturally could have no interest for the ordinary person. And as to universality of acceptance, the more nearly possible it is to distribute a single program across the country the more nearly will advertisers, if that type of exploitation remains, be able to achieve the highest possible consumer interest. Also, the more nearly will a political candidate be able to reach all the people simultaneously. And the more nearly will all the people be able to see, as it happens, some major news event.

And here we stumble again upon the difficulties of program selection. Mr. Skinner reminds us of the clash between Messrs. Jolliffe and Jewett, when he points out:

> It must be assumed that if a given channel is assigned in Boston, that channel cannot be assigned to any other center nearer than Philadelphia, and any channel assigned in New York cannot be assigned again any nearer than Baltimore or in Buffalo. Similarly, any channel assigned in Cleveland probably cannot be assigned in Toledo, Akron, Youngstown, Buffalo or Detroit.
> It is not likely, at least in the early days of broadcasting, that adjacent television channels could be assigned in the same city, because of probable interference.[3]

Nationwide service must be the goal of television, Mr. Skinner feels, even though that really results in exclusive operations in a given locality by one licensee. This condition we know must often be the case, for sets designed to receive programs broadcast on one frequency and definition cannot make coherent the programs sent out on another. Here he falls into a contradiction.

Mr. Skinner admits that it will be difficult enough even to distribute a single program on a nationwide basis, but insists, nevertheless, that competition and the public interest be served by offering the residents of a single community at least two television programs from which to choose. How to resolve this conflict with his principle of imperative nationwide service, Mr. Skinner does not say. The public must naturally bear the cost of distribution, the RMA feels. The history of cost is enlightening. Sound radio sets, between 1924 and 1929, cost on an average of one hundred and ninety dollars, only to drop in price to present levels after technology (and lawsuits) had provided higher standards of efficiency and lower costs of operation through experimentation.

Finally, Mr. Skinner states:

In the opinion of RMA, the Federal Communications Commission has in television a great opportunity and a great responsibility. Here is an impartial body, and with no interest to serve but the public interest.

The public is already aware of television. The public not only wants television, but it expects television, and it seems to be getting somewhat impatient over the long time it is taking to work out.[4]

All these are valid words, and significant. They convey as much of warning as of invitation to the Communications Commission. The public and Mr. Skinner are sitting in to see that the commission does not forget that it has no interest to serve but the public interest.

Which brings us to an examination of the commission. Let us re-emphasize the importance of this group's position. It must set standards of performance which will have infinitely ramified effects. Here are just two examples of the repercussions which may be expected from its decision:

238   TELEVISION

The television facsimile service may lead ultimately to a decision on whether radio or newspaper interests will control dissemination of news. . . .

Facsimile will broadcast a full newspaper, banish newsboys, presses, delivery systems, make dot and dash telegraph as obsolete as the pony express, by visual transmission of information, weather maps . . .[5]

Labor recognizes the threat inherent in such an event. William Green, president of the American Federation of Labor, observes "radio is more important to the public welfare than the newspapers."

And in the second example:

Television appears to be a rich and fluid medium, and writers and directors, especially, might be eager to see what they could do with it.

At this point, however, some cold realities of engineering and economics intrude themselves. A television channel is an exceedingly costly thing, running into hundreds of thousands.

Further, the great plaint that radio uses up literary material too fast (one broadcast on one evening and the manuscript is finished for all time) is as nothing compared to what television will do to stage settings.

A theatrical producer, planning a season's run for his play, can invest in substantial settings, but what would happen if he had to change his play and his sets not only every night but several times in a single night? Again, the cost approaches the fantastic. . . .

In this dilemma, the Farnsworth studios are working on an ingenious solution based on the use of miniature sets.[6]

But suppose all these problems of frequency allocation and program detail are settled. We cannot resist the indication of another, a problem supposedly settled at the outset, concerning technical quality of operation itself. This is a bit of news from a laboratory most people believe quiescent. Nothing could more sharply and dramatically remind us of the ceaseless dilemma with which the regulatory commission is confronted, as it seeks simultaneously to protect and foster public interest and private enterprise, than:

I would like at this time to advise the Federal Communications Commission that we have designed and patented a mechanical system [of television operation] in which a new and revolutionary principle is involved.

The principle is so radically different from that of any other system heretofore used that it would not be possible to adapt any of the present existing methods of inter-laced scanning to this system, although it does utilize inter-laced scanning.

This receiver is capable of projecting a three foot square picture with a definition of two million picture elements, and although this definition is considerably higher than that contemplated by some of the present companies, it has been demonstrated as commercially practicable. . . .[7]

By now the reader realizes how sensational a statement this can be, if true. Out of the tortures and tribulations of sound radio, RCA and the Bell system have laboriously built themselves to powerful positions in communications. They have won after battle and compromise. Great sums have been paid out in lawyers' fees and other costs to develop patent positions giving them dominance. And dominance, insofar as RCA is concerned, is predicated upon the cathode ray scanner, the iconoscope. And though the Bell system is not promoting any particular type of scanner, it is into this field deep enough to protect its stake in the coaxial cable, which it hopes to force RCA into using, in preference to the relay, point-to-point booster of radio waves.

Who challenges the giants, then? It is R. D. LaMert. And who is he? R. D. LaMert speaks for the De Forest Television Company, of Hollywood, California. This is the truest sort of drama. Lee De Forest, who is called the "father of radio," the great elder, the inventor who got a pittance and could not keep it, returns to the wars with a new invention, threatening to control the new art.

De Forest can say, without any fear of challenge, that RCA and the Bell system are nothing more than the corporate expressions of his own genius. By inventing the three element thermionic valve, he made them. By inventing the radical mechanical television scanner, will he unmake them?

This question can be answered in one of two historical ways. De Forest is old, and he is almost alone. He can fight at this late day through the Supreme Court of the United States, but is unlikely to do so against these two aggregations of power. He can surrender and sell his wonderful new device to one of the two survivors from the early battles. He comes close to holding a balance of power over his ancient enemies and partners. The Supreme Court may yet decide between him and them.

21. The Seven Wise Men

TOWARD WHICH SIDE IN THE GIANTS' STRUGGLE WILL DE FORest throw his new weapon? We know that the very nature of television is such that no individual alone can construct a system and operate it. He may receive considerable income from patent royalties once production is instituted, but nothing less than a fortune running into millions of dollars could actually organize the going concern. De Forest is a man of temper and strong feelings. Undoubtedly he bears no love for RCA, after his experience with that organization in the thermionic valve business. Perhaps he feels a grudge against the Bell system because of that transaction of long ago concerning the original patent which brought such wealth and power to the corporation and so little to him.

Indeed, this real "father of radio" has recently begun to exhibit a profound disgust with the whole business, and has stated that unless reforms are instituted in radio he would have bitter cause to regret that he ever brought "this American Frankenstein" into existence. But how can such reform be effected? De Forest, Farnsworth, Zworykin, Alexanderson—none of these could possibly conceive of a mechanical system of administration. Equitable and intelligent operation of communications depends upon human beings. Responsibility for such rests exclusively with the Federal Communications Commission; and it is from that body primarily that any reforms must come. If the commission declines, then Congress, which created it, has the legal and moral mandate to act for the general public. Who, then, are the commissioners? What is their attitude toward their task?

We have already reviewed the growth of the law which is supposed to predicate their actions. We realize the size and strength of the organizations which they are supposed to regulate. Now let us see specifically the kind of men they are and the kind of decisions they are inclined to make. Membership of the commission has been wont to change rapidly, so there is first of all to be acknowledged a complete absence of any definite corpus juris or consistent line of opinion in their findings. Superficially this may appear an extremely bad trait, but in view of the fact that the only constant in radio is change, an attitude of flexibility probably is the soundest that could be adopted by any intelligent men. Whether the commission shows a broadness and wisdom indicated as necessary by the immensity and importance of the subject is a matter the reader must estimate for himself. And so, let us summarize the public facts known about each commissioner now sitting, and examine some decisions the commission (not always with all the present members upon it) has made in important instances.

Frank R. McNinch, chairman. He is a North Carolinian, a lawyer, and a member of the Democratic party who supported Herbert Hoover for the Presidency in 1928. He came to the Communications Commission in 1937 by special order of President Roosevelt on a one-year leave of absence from his regular assignment as Chairman of the Federal Power Commission, with a plainly labeled order to straighten out conditions that were then the subject of common gossip. Shortly afterward there developed a theory within the Capital that the Communications Commission would be abolished, ultimately, and its duties consolidated with those of the Power Commission or the Department of Commerce. A dry-mannered, cautious-spoken man, McNinch is generally feared as an uncompromising federal bureaucrat but not as a rampant reformer. The radio industry has approached him with caution and unction, and is not particularly satisfied with his statement of opposition to congressional inquiry into the communications industry or the Communications Commission. McNinch has said in unvarnished language that he has found the commission and the industry in a state most charitably described as "unsatisfactory," and has indicated his intention to act against monopoly and indecent programs.

Thad H. Brown, Republican, of Ohio, also an attorney. Mr. Brown is a carry-over from the Federal Radio Commission. He was distinguished as the center of controversy in the WNYC case already mentioned, in which he refused to excuse himself from hearing in the case after challenge, and was sustained in his position by the U. S. Court of Appeals for the District of Columbia.

Eugene O. Sykes, Democrat, of Mississippi; former state circuit court judge. Mr. Sykes, like Mr. Brown, was on the Federal Radio Commission. When this gentleman came before the Senate Committee on Interstate Commerce for confirmation, in 1935, his fitness for continuance in office was challenged by Senator Theodore Gilmore Bilbo of his own state. Bilbo accused Sykes of having used his influence as a member and acting chairman of the commission, during the 1934 primary campaigns of the Democratic party in Mississippi, to prevent his, Bilbo's, election. Among other things, he said that Sykes had lent the color of his authority to a plan linking together several radio stations broadcasting speeches without charge in opposition to Bilbo. This Sykes and all the people accused with him promptly denied, and Bilbo was never able to make a clearcut statement of proof against them, though three out of the four stations involved admitted carrying speeches by Bilbo's opponent free of charge. The manager of the fourth admitted use of Sykes' name in arranging the broadcast by the others.

Bilbo's second charge:

I now invite your attention to a telegram addressed to President Franklin D. Roosevelt, dated November 19, 1934, and sent by George Llewellyn, of Atlanta, Georgia, formerly assistant supervisor of radio, Atlanta, Georgia, making special reference to Judge Sykes as being involved in certain charges that he [Llewellyn] had made to the Department of Justice agent....[1]

The Llewellyn case is a story of procedure on a par with the KNX case of Los Angeles in which transfer to another assignment came to the agent of the commission who reported some forty alleged violations of law by the licensee station operator. Sykes, it ought to be stated at the outset, was shown conclusively to have had no connection with the Llewellyn matter. He was in Europe during its developments. But that does not dispose of the facts in the case, which are admirably summarized in a cross-examination of

THE SEVEN WISE MEN 245

Commissioner Brown by Senator Burton K. Wheeler, of Montana, chairman of the Committee of the Senate on Interstate Commerce:

The Chairman. You kicked the one boy [Llewellyn] out of the service because of the fact that he told you of misconduct on the part of his superior officer [Van Nostrand], yet you reinstated Van Nostrand and permitted him to resign; and then later, he is permitted to pass upon regulations as to whether or not these various broadcasting stations, among them the one out of which he got money. This Van Nostrand is permitted to regulate these stations.

Commissioner Brown. Senator Wheeler, our inspectors in Atlanta determine that in making their reports to the Engineering Department of the Commission. Major Van Nostrand, if he is hired by a private station, may make reports, but so far as I know, they would have no more consideration than a report made by anybody else outside.

The Chairman. But your own secretary refers a broadcasting station to a private company. [For engineering test service to be reported to the Commission, which maintains its own test service.]

Commissioner Brown. I think that is entirely wrong, then.

The Chairman. Nevertheless, that is going on?

Commissioner Brown. This is the first time it has been called to my attention.

The Chairman. It seems to me that the Commissioners are not paying attention to their duties when they do not know what is going on from their own secretary and their own engineers in charge.

And that [Llewellyn incident] is not the only case.

This record here is filled with them. This record also shows that previous to the time that Tifton sold his station, that pressure was being put on by Van Nostrand, constantly harassing him, saying that he was violating the rules, prior to the time when he sold out, and then it shows that Van Nostrand was urging him to sell his station and at that time that he had an application in and wanted to build a better station and so forth; that pressure was being put on him by Van Nostrand to sell out.[2]

This statement from the records of the Communications Commission by Senator Wheeler had a sudatory effect upon the commission, which promptly reinstated Llewellyn and let Van Nostrand drift. It is also a classic summary of the sort of affair which has led to resolutions urging the most thorough examination into all the commission's records by Congress. But let us continue examination of the commission personnel.

George Henry Payne, Republican, author, editor, and former newspaper correspondent, of New York City. He had no experience in communications prior to his appointment on the commission, but has an ample stock of opinion on it which has been set forth in a foreword to this book. Mr. Payne has stated to the authors that in fifteen years' service on the board of tax appeals in New York he saw nothing to compare with the Federal Communications Commission. He has been accused by publicists of the broadcasting industry of being a demagogic politician, and has responded by actions in libel. Mr. Payne, as we have already indicated, holds that radio is being ill-used. He has not always found himself in agreement with his fellow officers on matters of procedure.

On one occasion he was debarred by them from a part in adjudicating charges of improper conduct brought against an attorney practicing before the commission. The lawyer, Paul M. Segal, happened to be counsel for the respondents to Payne's libel suit. He was also the subject of an extensive investigation by a sub-committee of the commission of which Payne was chairman. When Segal came on for trial, on the charges of improper conduct, he pleaded that Payne was prejudiced, that he, Segal, had done nothing wrong, and that his methods were common practice before the commission.

The commission voted five to one against Payne's sitting on the Segal trial, and used the practice of the Federal courts to buttress its position. It ignored the ruling of the Court of Appeals in the case of Commissioner Brown that a member's position is unassailable except by impeachment in the House of Representatives or removal for cause by the President of the United States. After a lengthy trial, the commission found Segal guilty of improper practice in that he filed applications for station licenses and failed to reveal the identity of the true applicant. He was suspended from practice before the commission for sixty days. No inquiry was made public concerning his charge that his action was common procedure before the commission.

Paul A. Walker, Democrat, of Oklahoma. He is the unchallenged authority of the commission on matters of telephonic public service. He came to the Communications Commission from chairmanship of the Oklahoma State Corporation Commission, which regulates public utilities in that jurisdiction. Walker directed the Communications Commission's two and a half year investigation of the American Telephone and Telegraph Company. This inquiry was the most expensive and detailed ever undertaken by any governmental agency. Its direct bills are in excess of $1,500,000. Just what it has accomplished toward establishing concrete regulation of the Bell system there is no way of telling, as yet. The Bell system has flourished free and clear of Federal control for so many years that its regulation now is a lengthy, intricate task. Not until passage of the 1934 communications act was it under direct supervision of any agency. The investigation just concluded appears rather to have established a volume of reference for further study and investigation of telephony than to have evolved any immediately applicable standard policy of regulation, but it is directly credited with having brought about reductions in excess of twenty-two million dollars in long distance tolls. It is an axiom of the public utilities world that the Bell system is too smart and too powerful for any state public utilities commission to regulate it. Whether the same shall be said of its relations with the Federal Government depends upon how the Walker inquiry is finally resolved upon by Congress.

Norman C. Case, Republican, three times Governor of Rhode Island, and an attorney, but with no experience in radio law. Case has not been noted either for vigilance or vigor. However, he made one comment during the proceedings of the Senate committee testing his fitness for office which indicates his realization of the general problem before the commission:

You cannot stop, Senator, the improvement of the technical development; and if the radio point-to-point communication is a better service you are naturally going into that service; you cannot stop the evolutionary advance.

It is a fact that radio does give this point-to-point service and that the wires are becoming somewhat obsolete.[3]

Will Mr. Case remember that declaration when the final resolution comes between the Bell system's coaxial cable and the radio spectrum which RCA visions as its own pearl-sprinkled oyster?

Teunis Algiers Monterey Craven, only member of the commission who is also a qualified radio technician, is deserving of a special analysis. Mr. Craven's career is an interesting example of the play of personal interests between public office and private institutions which has characterized the radio regulatory commissions since their inception. Commissioners, legal counsel, and engineers have found themselves first on one side of the bench in hearings for stations licenses, and then the other.

Craven testified to the Senate committee that he became a student of electronics while a midshipman at the United States Naval Academy, and followed a natural interest in the subject as an officer in Naval Communications. In 1930 he resigned from the service of his country, and offered his technical ability to the highest bidder as a consulting engineer on radio frequencies. It became his habit to testify before the commission on behalf of applicants who might retain him for a cash consideration. Mr. Craven disclosed on cross-examination that he adopted this practice after having served with the Radio Commission on detached service from the United States Navy and found how it operated.

About the middle of November, 1935, he said, while he was at the peak of his private career, Chairman Anning Prall asked him to return to the service of the Government as chief engineer of the commission. Craven's testimony to the Senate committee was that he did not wish to return to the Government service but that Chairman Prall insisted. "He stated," Craven said, "that the President's views were that I had been educated by the Naval Academy at the expense of the Government and that the Government had afforded me an opportunity to develop myself in the field of communications, and that as a result of this training which had been afforded me by the Government, I had been able to establish myself as one of the few consulting radio engineers in the world."

A call to duty by the President of the United States, Craven said, left no alternative.

I accordingly made a great personal sacrifice in order to comply with the request of the President. I divested myself of all interests which I had in every station.

I also disposed of my practice. ... At that time the practice was worth approximately $50,000 a year gross to me, meaning possibly $30,000 net. I disposed of it ... for $15,000, the highest figure which could be obtained on short notice.

A few more questions brought declarations of opposition to monopoly in communications, of belief in governmental regulation, of a promise to "be loyal to the President of the United States," and to follow his committee chairman's lead on everything lawful and "right." The Federal Communications Commission's only avowed engineering member was then recommended by the Senate committee for confirmation. He was so confirmed, and represents the people of the United States as their chief technician, it might be said, in disposing the fates of stockholders and guiding the uses of the electron for radio communication.

As an engineer he should take on considerably more importance than do the legalists on the commission, for the development of adequate radio facilities is primarily a technical problem. How, then, does he view radio? While chief engineer, he undertook a study of the social and economic aspects of radio, the first of its kind. "Social" he defined as service to the people of the United States, including the extent to which broadcasting assists in the development of national, community, and individual well-being. "Economic" was held to be the aspect of radio as a business. In general, it might be said that his approach to radio, the dynamic, constantly changing factor of human affairs, was in the mood of one analyzing some relatively static enterprise such as the cotton goods industry, the retail department store trade, or the production of brick tile.

Consistently, Mr. Craven has declared himself for competition, business competition, in radio. Social service, he appears to conceive, consists of two or more radio signals between which the customer can choose in a single market area. To that end he made an engineering recommendation that the portion of the spectrum now encompassing sound broadcasting be expanded by 100 kilocycles and then redivided into one hundred separate frequencies instead of the present ninety. Instead of four classes of transmitters he would have six, with the object of penetrating market areas on a more nearly even basis. The social aspects of radio, Craven considers, are somehow inextricably insured by continuance of the present principle of commercial marketing. He determined that of the 16,598 cities and towns in the continental United States, six hundred and fourteen with population in excess of ten thousand are without radio stations. However, only one hundred and eleven fail to receive programs on a reasonably satisfactory technical basis.

To what degree the technical efficiency of operation would be improved by Craven's plan of reallocation would appear to be far from a settled matter. Should these one hundred and eleven towns have radio stations to the detriment of present service? Should superpower broadcasting which reaches rural areas and others of relatively sparse population be reduced in the interest of superior competitive marketing activities in more densely populated areas? In sum it should be stated that Craven appears positively and finally committed to what he calls the "American system" of radio service financed by revenue from merchandising, but on the basis of commercial competition. He appears to pursue that principle even into other divisions of communication than broadcasting proper. In January, 1938, when the commission had before it a petition from Western Union, Postal Telegraph, and the RCA and Mackay systems for authority to increase rates and charges for message service, Craven seized the occasion to dissent from the opinion of his fellow commissioners that hearings should be ordered; and he declared himself first for a more fundamental study of the whole communications structure with a view to searching out possible plans of better competition.[4]

The question of competition, it is well recognized, is one vital to pure business enterprise. Competition may, but not necessarily must, serve the public interest, necessity, and convenience—to protect which is the primary responsibility of the commission. Judging from the record, the commission has no clearcut idea of when and how to invoke competitive principles. On one occasion, the Mackay Radio and

Telegraph Company entered a petition for permission to install service between the United States and Norway in competition with RCA, which then had an exclusive franchise for radio operations in that field. The nature of this competition may be estimated from the fact that in 1935 eighty-eight per cent of the westbound, and sixty-two per cent of the eastbound, communications traffic between Norway and the United States was by radio. The Mackay organization, with heavy investments in cables, sought to shift its principle of operations with the technological and business trend, offering to match RCA's radio communications as to price, speed of operations, and classes of service; but it was denied the right on the ground that to do so would be to eliminate the cable systems almost entirely without substantial improvement in competitive aspects. The argument that radio operation would strengthen Mackay for better competition with RCA than the cables were then offering did not appear to impress the commissioners as a business fact or possibility. This decision was rendered on April 24, 1937. Its effect was to stagger any operators of cable systems who had hoped to follow the Mackay plan of transferring their competition with RCA from cable networks, which are obsolescent, to radio, which is improving constantly in technical performance and in business efficiency.[5]

22. Public Policy

AND SO WE COME TO THE CONCLUDING QUESTION, "WHAT shall we do about television?" There can be no challenge to the use of "we," for the declared policy of the Republic is that interstate commerce in electrical communication, whether wire or wireless, shall proceed only in the public interest, necessity, or convenience. We have something of an understanding of the basic principles upon which radio technology is founded. We know the history of custom and law developed with the changes in the technology.

There can be no challenge to the statement that the future of politics and social order and the future of television will follow parallel courses. Sound radio has already precipitated the fall and rise of governments. On one occasion (March 4, 1933) it stabilized a great nation gone hysterical. Nobody can recall the first inaugural address of President Franklin Delano Roosevelt ("The only thing we have to fear is fear itself") and deny the social importance of radio communication.

But what is the status of administration? Some portions of the geographic United States do not receive enough service; others receive too much. Sound broadcasting is contaminated to the extent that it is almost always thought of by the common listener as an adjunct of commercial advertising. The financial organizations most closely connected with broadcasting work toward monopoly and suppress technology in the interests of business stability. Programs are offered not always with the highest motives, not chiefly for entertainment and instruction, but primarily for the purpose of propagating sales of goods. Licenses fall too many times into the wrong hands. These are commonly uttered criticisms of the sound radio broadcasting industry. They should, however, be directed not against the commercial interests but exclusively to the Federal Communications Commission. For the commission is the people's representative in a convergence of operators who have been authorized to function solely in the public interest, necessity, and convenience. Yet the people's representative has failed to declare positively just what the public interest, necessity, and convenience encompass.

If there are too many radio stations in some sections of the country and not enough in another, has not the commission power to remedy? Obviously. The records of its proceedings are filled with complaints by radio interests that the "spectrum is too crowded." Yet the number of stations increases. The commission revokes only about two operating licenses a year, and it allows the net total to increase. Yet it does not distribute individual stations to the best geographic advantage.

If the operators devote themselves more to selling timespace for advertising and less to good entertainment and education which bring in no cash returns, has not the commission power to revoke licenses in cases of neglect to the public interest? And can it not do the same when political utterances guaranteed under the Bill of Rights are censored? Can it not do the same when the canons of decency and good manners are violated? Can it not do the same when any matter of public interest, pertaining either to the quality of goods advertised or the nature of news events, is misrepresented or ignored? We know that it can do all of these things, for it has absolute powers of determination as to what is encompassed by the term, "public interest, necessity, or convenience."

And we know that if it does not do its duty, the fault is our own. It is illogical and unfair to expect the Radio Corporation of America meekly to surrender its monopolistic ambitions, or the Bell system to write off as a financial loss its great network of telephone wires in the face of technological change. It is absurd to think that either of these institutions, if it can buy up the patents and devices of some inventor like Philo Farnsworth or Lee De Forest, will not use these to its own profit, regardless of the effect upon the public or upon competing business groups.

Sound radio has demonstrated that the communications business is conducted on the basis of a titanic struggle to protect huge investments and attain great profits. It has demonstrated that the very essence of use in a communications instrument demands an attempt by the exploiter to monopolize the field of operations. The more nearly universal the acceptance of a device is, the more valuable it becomes not merely to the exploiter but to the user. A telephone that connects the international business man with Paris, Bangkok, or Berlin is more important to him than one reaching only into the next county.

We have not dealt in detail with many criticisms of the

Communications Commission which, however valid, are relatively minor. Analysts of administrative government hold that it has a faulty structure. A seven-man body charged with both executive and judicial functions tends to become a debating society, they hold. A similar statement might be made about the Supreme Court of the United States, unquestionably the final source of power in our constitutional government. The tendency to debate rather than execute duties would seem to depend upon individual will rather than mere communal session.

The commission has been made the butt of political manipulation and patronage abuses. This is one of the most common criticisms of all, but it is also one of the weakest. Political manipulation and patronage abuse can be carried on under any form of government; and so they are. In a democratic republic they proceed exclusively by sufferance of the citizenry. If the public objects to maladministration, it knows the remedy it has itself provided.

There are some criticisms of the communications law which must be considered, too. Investors in radio financing and operators of radio systems declare that the licensing provisions leave them in a state of nervous apprehension. A man has no tenure, they say, no certainty that he will be allowed to continue operations for more than six months. They ask licenses for not less than five years at a time, and also some guarantee that property rights will be protected in the event renewal is denied. The history of administration offers no tangible basis for such fears. Licenses generally are revoked only in the most flagrant instances of abuse or financial inability to perform. None of the great broadcasting chains or their stations has suffered such punishment yet.* License holders, except in rare instances of experimental operations, are not even required to carry the burden of proof that they serve the public interest when they come in for renewal of co-operating permits.

A five-year license obviously would insure the operator more adequately against the perils of technological change; but it would also operate to restrain advance in public service.

There is a strong congressional movement for limiting the licensing rights of the commission in such a way as to prohibit joint ownership of radio stations and newspapers. The theory appears to be that radio and newspaper should compete for the news, and offer contrasting editorial opinions upon the events of the day. Such a condition, it is held, would insure against monopoly of information and distortion of the public mind. Here, again, the remedy lies not in legislation but in public action. If the law allows development of great chains of commercial broadcasting stations of the sort typified by National Broadcasting Company, and continues the present practice of granting broadcasting licenses to set up facsimile systems of broadcasting, the printing press newspaper may find itself no longer an important instrument of competition. It, like the legitimate theater in relation to motion pictures, tends to become just a sort of testing service.

Radio is being monopolized at the source, not at the outlet. The final licensee operator of a station, whoever he may

* The only indication of a change in policy toward these was made by Chairman McNinch after the commission held that NBC had violated the canons of decency in the Mae West broadcast. He said that case would be considered when the fifty-nine stations using the program in question applied for license renewals thereafter.

be, is at the mercy of the Federal Government and the great organizations typified by the Bell system and RCA. His dilemma is in no way solved simply by making him a radio operator-groceryman, rather than a radio operator-publisher. And the newspaper publisher, presumably skilled in the difficult art of satisfying a great majority of the community as to entertainment and information, possibly is more to be trusted than the grocer, the banker, or the insurance executive in developing good radio program policy. No evidence has been offered that publishers have been or are worse or better than the average station licensees. They just go into radio as rapidly as possible, simply as a hedge against the day when facsimile may put the press out of business.

The spirit of this proposed limitation upon them might better be preserved by granting licenses only to bona fide residents of communities in which stations are operated, and upon proof that the public is being served in the best manner possible. By this means none of the values in nationwide broadcasting of single programs would be lost, for chain networks would continue but local interests would be more likely to dominate in editorial handling of news and politics.

Inescapably, as one ponders these problems raised by sound radio and shadowing the future of television, one finds there are three basic questions:

First, shall radio, sight and sound, continue as unrestrained, untaxed, private enterprise under the present system of licensing?

Second, shall it become a closely regulated public utility,

with fixed rates and tariffs comparable to the telephone industry?

Third, shall it be liquidated as a private enterprise and operated exclusively by the Government?

Let us keep in mind the historical background as we examine these three, seeking to analyze the future of television.

Television, structurally, is a synthesis of communicative forms. It combines sight and sound. Operatively, it is as the lock and key. The whole function of the manufacturer is to serve the holders of the lock and the holders of the key, the transmitters and the receivers. Television is a medium of information and entertainment for the control of which a terrific struggle is being waged. It is a medium also for acquiring great profits both in money and in power. It is coming into ordinary use slowly; but if the history of invention is not to be denied, television will in time become as common as the sound radio is today. It is expensive now, but ultimately the price will meet the market demand because technological advance and change have been found to achieve such results, however incidentally.

Technology, in itself, guarantees nothing save change. When inventors have succeeded in developing adequate standards of performance they have done their job in life. It is no duty of theirs to be concerned with bankruptcy courts, frenzied investors, price structures, vested interests, or the trading philosophy of the Bell systems. When De Forest discovered that a grid would modulate the flow of electrons, he had merely to put the grid in its proper place; and so it was with Edison, Hertz, Marconi, Alexanderson, Steinmetz. How easy, compared to the plight of the business man, the lawyer, the stockholder, and all those who would regulate the economy in which television must find a place!

These inventions, in overcoming the problems of techniques, create engines that strain and sometimes destroy the economic structure expected to accommodate them. And the imperative of accommodation is one the lawyers, the business men, the economists and commissioners, try as they will, cannot deny. A telephone is invented, and soon all must have it; and so too with the electric light and the automobile. Government fiat, suppression by vested interest, and the activity of those who stand to lose by the development of an invention may delay and harass its progress, but if historical precedent is to have any meaning, we have to admit the invention is accepted finally. Unfortunately for our peace of mind the familiar institutions of profit, free private enterprise, free price, private property, which have regulated our economy are inadequate to effect a painless acceptance of the new state of affairs when invoked in their pure forms. They were developed in a handicraft economy, and we find them unable to function freely in conjunction with the highly mechanized, integrated machine technique. But after the clash of inventions and established institutions, modifications in the character of control always seem to take place, for an invention is used as it is conceived—or it is not used. That is, a machine is a machine and nothing more. It performs one task and that inflexibly. It is control that must be flexible and adaptable. And flexible control has always managed to make a place for new machines so far. How will adjustment come about in the case of television?

Upon the resolution of our three basic questions rests the fate of industries with investments in the billions of dollars: the future of communications, the test of government regulation, the radio industry's subdivisions of transmission, manufacturing, and advertising; the motion picture industry with its technicians, furriers, heroes, and heroines; the telephone monopoly and its seven hundred thousand-odd stockholders and all those dependent upon them.

Whoever gains the initial advantage of pre-emption will have a major power over the many others who, in order to continue existence, must have a part in television. But being first has its perils; and this is a warning to sound radio in its fight for control. The initial investor must cope with the capriciousness of technological change; the economic yardsticks of profit and loss; the Federal Communications Commission's amorphous definition of public interest, convenience, and necessity; and a mystical winnowing of all these in the flailing chamber of the United States Supreme Court.

Will the first entrepreneur in television serve in the manner of the male bee, simply to fructify and die? There is strong chance of this, for television requires heavy investment for plant, personnel and operating material. Errors in judgment, therefore, will be penalized severely. Uncertain, faltering regulation will be fatal to all concerned; industry, the public, the Government, the economy in general. This ought to be obvious, but seems not to be. Unquestionably capital stands ready to bring out television. But upon investment there must be a return. If television is to be privately operated, those interested must recognize that regulation is a technical imperative, and that the healthy condition of the art requires regulation to be stringent and honest. For, to attain a measure of the precious stability necessary for successful commercial operation, the interested parties within the industry must know on more than a special privilege basis where they stand in regard to regulatory administration and policy.

Special privilege is an ephemeral thing. What can be achieved today may tomorrow be passed on to a richer, more influential interest. Operating on a six months' license, or even a three or five year license, the entrepreneurs are entitled to know what definition is going to be applied to that much too mysterious phrase, "public interest, convenience, or necessity." By the same token, they must consider seriously the elevation of program standards, the problem of balancing between "editorial selection" and censorship. An aroused public opinion is at times very costly to investment. If the public has been lax to the implications and operations of aural radio, that does not mean it will be lax with television. The growth of "Legions of Decency" and the rumblings in Congress are indicative.

The Federal Communications Commission will be a potent factor for good or evil, of course. Its history in the radio broadcast fields is well known. What will be its position in television? It will decide the problems of allocation on the spectrum, fix standards and interpret the institution of law as it relates to the art. The question of listener and viewer interest will be posed to it. What will be the rights of the receiver of programs if the construction of a steel building in his vicinity interferes with technical reception? What if a doctor's diathermy machine conflicts with a program? But most pressing and immediate are the definition of the public interest, convenience, and necessity as applied to the granting of licenses, the allocations on the spectrum, and the fixing of standards.

In addition to the spectrum problems, there are standards of performance to be fixed. Transmission and reception are reciprocally dependent. One type of transmission technique requires a similar type of reception technique. And there are many inventors, each claiming that his instrument should set the standards.

But the problems of monopoly, place on the spectrum, and technical standards are soluble by simple fiat, or economic strength. They are just empirical tests of quality. They are not imponderables, merely problems involving exercise of choice. But there is another problem here that tortures the sleep of the business man: once television is out, can it be made to pay profits and remain stable in technical development? There, somewhat simplified, is the crux of the present impasse. Adapting the electron to create pictures is one thing; making it profitable is another. Some pretty problems have resulted from the attempt to squeeze television within the institutional framework of sound radio which operates on the principle of selling sets to the consumer and charging the cost of programs to the advertisers. In television the consumer can still buy his set, if the price is right. Some have attempted to pose this as a focal issue. Actually it is not, for under the system of mass production the price of sets undoubtedly can be brought down to reasonable levels. It is a good risk to say the audience is ready.

But the cost of programs is really maddening to those who would share in the television harvest. To keep it within the framework of sound radio this cost should be borne by the advertiser. Can he do his part? Experience has demonstrated that the cost of programs has been reasonable in relation to advertising potentials. This system apparently has been profitable to radio men and advertisers alike even at the price of $20,000 for one hour's entertainment on a national chain program. This is what Mr. Sarnoff and Mr. Paley call the "American Way."

And in television they would also like to operate in the "American Way," of course, but television is very unpatriotic and up to the present it doesn't seem able to conform. For one thing, as has been noted, its character is embarrassingly monopolistic, setting it counter to traditional competition. Yet more embarrassing is the search for someone to bear the cost of programs, most of which, it is well established, will be in the form of film motion pictures. A motion picture feature giving an hour's entertainment costs from $350,000 to $1,000,000 and sometimes more. What advertiser can bear this cost for an hour of television?

But should it be found that the cost of film can be circumvented or solved, another difficulty arises. Will the eye accept advertising in television? The experience of motion pictures has demonstrated that attempts in this direction are dismal failures. In 1937, for instance, in a city in Missouri, groups of "movie" patrons took it upon themselves to boo and shout catcalls when advertising appeared on the screen. This appears a very bad omen—but do not underestimate the advertisers and radio men. For instance, it has already been found that by reducing the size of the film from the standard thirty-five millimeters to sixteen, substantial savings result. By the use of miniature sets in perspective against neutral backgrounds, more economy is accomplished. Those vested with the guidance of radio and advertising are resourceful and ingenious. And they are working frantically to solve their difficulties.

One fact remains: however much they seek to work separately, still they must come together. Progress in engineering is and always has been a cumulative expression of all technical information. The engineers of RCA do not work isolated from the world any more than their competitors do. They exchange information, rush to their laboratories, try to accomplish new results ahead of the other fellow. So do stage designers, managers of performance, financiers, and program directors. Then, the new analysis supposedly achieved, the triumphant one demands of his government protection in the form of patent, copyright, judgment for damages in plagiarism. He seeks external help because alone he is helpless.

Clearly, then, television cannot escape government dominance, however much effort is expended to make it conform to the principle of free, private enterprise. It is simply a matter of how much dominance. There are business men who want to remove television entirely from the influence of the traditional sound radio technique. They are especially concerned about the effects of the synthesis upon motion picture exhibitors and their interests. Why not set up television as a public utility? Thus do we find ourselves sliding from the premise of free competition and free profits to limited and guaranteed income in return for guaranteed public service on a common carrier basis, comparable to telephony.

This, it is claimed, would make possible more effective

regulation, and permit the economic stability in television so notably lacking in sound radio. In addition, the problem of cost, now so difficult for those who would like the competitive advertising method, might be solved simply by charging service costs to the consumer on a utility rate basis. This proposal has been put most concisely by Robert Robins, executive secretary of the Society for the Protection of the Motion Picture Theatre, an organization of independent theater owners, radio set manufacturers, and other imperiled interests. Mr. Robins, appearing before the Informal Engineering Conference of the Federal Communications Commission, outlined a three point program.

Television service in its early stages, he held, must be confined to entertainment and educational purposes, such as the regular motion picture feature production, shorts, and newsreels; and television must be kept free of advertising. Furthermore, the programs must be a separate and distinct service, must be offered to the recipients on a service charge basis, and rates, rules, and regulations must be determined by a competent public body.[1]

His plan is very persuasive. The cost element is solved if consumers will pay. The difficulty here is that the possible consumers of television programs are the same people who now own radio sets. Their habits of thought have been so conditioned to receiving what appears on the surface as a free service paid for by the advertiser that the success of an attempt to burden the public with program cost directly is at least questionable. On the other hand, payment for electric light and telephone service is generally made without complaint. Television could be added to either of these without undue bother.

But other difficulties appear. If the transmission from studio to the home is to be brought through the means of the spectrum, the frequencies upon which television operates cannot be staked out or fused in to permit only qualified rate payers to enjoy their benefits. Individuals deserving to receive television programs but not anxious to pay the required fee will be tempted to buy or build receiving sets and "bootleg" programs into their homes as they have in England.

Strict competition would be absent from such a scheme. To allow television to develop on this basis in the hands of two or three public service companies suggests that those paying a fee to one transmitting company will be unable to receive the programs of other transmitting companies. If, to overcome this situation, the fees and rates are equally divided among the transmitting companies and the service of all broadcasters is open to all consumers, the incentive to supply better programs in order to attract more listeners is dulled. One just shares the current income and lets new business come on when it wills.

Why not throw the private interests out the window and put the whole matter in the hands of the Federal Government? The Government, it may be contended, could maintain the most elaborate sort of programs, extend the technical operating facilities over the widest physical areas, and continue development of the art and science, with a minimum of collision between interests. It is more able financially than any possible combination of private investors. And it is impersonal, has but one motive—the most superior possible service.

The Government could finance the installation of receiving equipment with a minimum of difficulty, and the whole expense could be met by the relatively simple process of taxation. But is that all?

The transient holders of public office adore power, and do not forego it without pain. What man ever willingly surrenders his seat among the mighty? If men were incorruptible, if ideals were never contaminated, if absolutism were really absolute and dependably moral, then the simple, efficient device of governmental production, distribution, and maintenance might serve for television and everything else. Do you think such a state of affairs is possible? And would you risk a civilization's future in a gamble for such perfection?

Appendix A

TELEVISION BROADCAST STATIONS 1937

LICENSEE AND LOCATION	CALL LETTERS	FREQUENCY KC OR GROUP	POWER VISUAL	AURAL
Columbia Broadcasting System, Inc., New York, N. Y.	W 2 X A X	B, C	50w	
Don Lee Broadcasting System, Los Angeles, Calif.	W 6 X A O	B, C	150w	150w
Farnsworth Television Incorporated of Pa., Springfield, Pa.	W 3 X P F	B, C	4kw	1kw (C. P. only)
First National Television, Incorporated, Kansas City, Mo.	W 9 X A L	B, C	300w	150w
General Television Corporation, Boston, Mass.	W 1 X G	B, C	500w	
The Journal Company, Milwaukee, Wisconsin	W 9 X D	B, C	500w	

APPENDIX A

LICENSEE AND LOCATION	CALL LETTERS	FREQUENCY KC OR GROUP	POWER VISUAL	AURAL
Kansas State College of Agriculture and Applied Science, Manhattan, Kansas	W9XAK	A	125w	125w
National Broadcasting Co., Inc., New York, N. Y.	W2XBS	B, C	12kw	15kw
Philco Radio & Television Corp., Philadelphia, Pa.	W3XE	B, C	10kw	10kw
Purdue University, West Lafayette, Ind.	W9XG	A	1500w	
Radio Pictures, Inc. Long Island City, N. Y.	W2XDR	B, C	1kw	500w
RCA Manufacturing Co., Inc., Portable (Bldg. #8 of Camden Plant)	W3XAD	D(124,000 to 130,000)	500w	500w
RCA Manufacturing Co., Inc., Camden, N. J.	W3XEP	B, C	30kw	30kw
RCA Manufacturing Co., Inc., Portable-Mobile	W10XX	B, C	50w	
The Sparks-Withington Company, Jackson, Mich.	W8XAN	B, C	100w	100w

APPENDIX A 273

LICENSEE AND LOCATION	CALL LETTERS	FREQUENCY KC OR GROUP	POWER VISUAL AURAL
University of Iowa, Iowa City, Iowa	W 9 X K	A	100W
University of Iowa, Iowa City, Iowa	W 9 X U I	B, C	100W
Dr. George W. Young, Minneapolis, Minn.	W 9 X A T	B, C	500W

GROUP A	GROUP B
2000 to 2100 kc	42,000 to 56,000 kc

GROUP C	GROUP D
60,000 to 86,000 kc	Any 6000 kc frequency band above 110,000 kc excluding 400,000 to 401,000 kc.

The low definition group (2 to 2.1 megacycles) is made up wholly of noncommercial licensees seeking to develop service for rural areas. In general these use mechanical scanning systems of about sixty line, twenty frame definition. Programs have been received as far as three thousand miles from transmitters.

The high definition operators (42 megacycles and up) are concentrating on intensely populated areas to which they expect to offer programs of an elaborate nature. In general, they have service areas of less than fifty miles radius, and use both mechanical and electronic type scanners, of four hundred line, thirty frame average definition.

Appendix B

NONPROFIT BROADCASTING STATIONS OPERATING IN THE UNITED STATES

CALL LETTERS	LICENSEE	LOCATION
KBPS	Benson Polytechnic School (R. T. Stephens, Agent)	Portland, Oregon
KFDY	South Dakota State College	Brookings, S. D.
KFGQ	Boone Biblical College	Boone, Iowa
KFKU	University of Kansas	Lawrence, Kansas
KFSG	Echo Park Evangelistic Assn.	Los Angeles, Calif.
KOAC	Oregon State Agricultural College	Corvallis, Oregon
KPOF	Pillar of Fire	Denver, Colorado
KPPC	Pasadena Presbyterian Church	Pasadena, Calif.
KSAC	Kansas State College of Agriculture and Applied Science	Manhattan, Kansas
KUSD	University of South Dakota	Vermillion, S. D.
KWLC	Luther College	Decorah, Iowa
KWSC	State College of Washington	Pullman, Wash.
WAWZ	Pillar of Fire	Zarephath, N. J.
WBAA	Purdue University	W. Lafayette, Ind.
WBBL	Grace Covenant Presbyterian Church	Richmond, Va.
WBBR	Peoples Pulpit Assn.	Brooklyn, N. Y.

APPENDIX B

CALL LETTERS	LICENSEE	LOCATION
WBIL*	Arde Bulova	New York, N. Y.
WCAD	St. Lawrence University	Canton, N. Y.
WCAL	St. Olaf College	Northfield, Minn.
WCAT	South Dakota State School of Mines	Rapid City, S. D.
WDAH	Tri-State Broadcasting Co., Inc.	El Paso, Texas
WEW	The St. Louis University	St. Louis, Mo.
WHA	University of Wisconsin	Madison, Wisc.
WILL	University of Illinois	Urbana, Ill.
WKAR	Michigan State College	E. Lansing, Mich.
WLB	University of Minnesota	Minneapolis, Minn.
WLBL	State of Wisconsin, Dept. of Agriculture and Markets	Stevens Point, Wisc.
WMBI	The Moody Bible Institute Radio Station	Chicago, Ill.
WMPC	First Methodist Protestant Church of Lapeer	Lapeer, Mich.
WNAD	University of Oklahoma	Norman, Oklahoma
WNYC	City of New York, Dept. of Plant and Structures	New York, N. Y.
WOI	Iowa State College of Agriculture and Mechanic Arts	Ames, Iowa
WOSU	Ohio State University	Columbus, Ohio
WQAN	The Scranton Times, E. J., Wm. R., Elizabeth R. & Edw. J. Lynett, Jr.	Scranton, Pa.
WSAJ	Grove City College	Grove City, Pa.
WSUI	State University of Iowa	Iowa City, Iowa
WSVS	Seneca Vocational High School	Buffalo, N. Y.
WTAW	Agricultural and Mechanical College of Texas	College Station, Texas

* This license has been assigned since last renewal of license was filed.

Bibliography

Aisberg, Eugene. *La Transmission des Images.* E. Cheron, Paris, 1930.
Albert, Arthur L. *Electrical Communication.* J. Wiley & Sons, Inc., New York, 1934.
American Academy of Political and Social Science. Annals. *Radio, the Fifth Estate.* Supplement, 1935.
Arnold, Frank A. *Broadcast Advertising, the Fourth Dimension.* Television Edition. Foreword by Alfred N. Goldsmith. J. Wiley & Sons, Inc., New York, 1933.
Baker, Thomas T. *Wireless Pictures and Television.* Constable & Company, Ltd., London, 1926.
Benson, Thomas W. *Fundamentals of Television.* Mancall Publishing Corp., New York, 1930.
Brindze, Ruth. *Not to be Broadcast.* The Vanguard Press, New York, 1937.
Brown, Frank J. *The Cable and Wireless Communications of the World.* Sir I. Pitman & Sons, Ltd., London, 1930.
Brunner, Edmund de Schweinitz. *Radio and the Farmer.* The Radio Institute of the Audible Arts, New York, 1935.
Buehler, E. C., editor. *American vs. British System of Radio Control.* H. W. Wilson Co., New York, 1933.
Butler, Hugh D. *The Radio Situation in Great Britain.* U. S. Gov't. Printing Office, Washington, D. C. U. S. Dept. of Commerce, Bureau of Foreign and Domestic Commerce, 1925.
Cameron, James R. *Radio and Television.* Cameron Publishing Co., Woodmont, Conn., 1933.

Camm, F. J. *Newnes Television and Short Wave Handbook.* G. Newnes, Ltd., London, 1935.

Cantrel, Hadley, and Allport, G. W. *The Psychology of Radio.* Harper & Bros., New York and London, 1935.

Chapman, E. H. *Wireless Today.* H. Milford, Oxford University Press, London, 1936.

Chapple, Harry J. B. *Popular Television.* Sir I. Pitman & Sons, Ltd., London, 1935.

Chapple, Harry J. B. *Television for the Amateur.* Foreword by J. L. Baird. Sir I. Pitman & Sons, Ltd., London, 1934.

Codel, Martin, editor. *Radio and Its Future.* Harper & Bros., New York, 1930.

Collins, A. F. *Experimental Television.* Lothrop, Lee & Shepard Co., Boston, 1932.

Crawley, Chetwode. *From Telegraphy to Television.* F. Warne & Co., Ltd., London and New York, 1931.

Dashiell, Benj. F. *The Beginner's Story of Radio.* The Radio Press, Inc., Cleveland, Ohio, 1935.

De Soto, Clinton B. *Two Hundred Meters and Down; the story of amateur radio.* The American Radio Relay League, Inc., West Hartford, Conn., 1936.

Dinsdale, Alfred. *First Principles of Television.* Chapman & Hall, Ltd., London, 1932.

Duncan, R. L. *Foundations of Radio.* J. Wiley & Sons, Inc., New York, 1931.

Dunlap, Orrin E. *The Outlook for Television.* Introduction by John Hays Hammond, Jr. Foreword by William S. Paley. Harper & Bros., New York and London, 1932.

Dunlap, Orrin E. *The Story of Radio.* The Dial Press, New York, 1935.

Eckhardt, G. H. *Electronic Television.* The Goodheart-Willcox Co., Inc., Chicago, 1936.

Felix, Edgar H. *Television, Its Methods and Uses.* McGraw-Hill Book Co., Inc., New York and London, 1931.

Floherty, John J. *On the Air; the story of radio.* Doubleday, Doran & Co., Garden City, N. Y., 1937.

BIBLIOGRAPHY 279

Franklin, Harold B. *Sound Motion Pictures, from the Laboratory to Their Presentation.* Doubleday, Doran & Co., Inc., Garden City, N. Y., 1929.

Gernsback, Sidney. *Radio Encyclopedia.* S. Gernsback Corp., New York, 1931.

Goldsmith, A. N., and Lescarboura, A. C. *This Theory Called Broadcasting.* Henry Holt & Co., New York, 1930.

Goode, Kenneth M. *What About Radio.* Harper & Bros., New York and London, 1937.

Great Britain. *Report of the Television Committee.* Presented by the Postmaster-General to Parliament by Command of His Majesty, January, 1935. H. M. Stationery Office, London, 1935.

Halloran, A. H. *Television with Cathode-Rays.* Pacific Radio Publishing Co., San Francisco, 1936.

Harlow, Alvin Fay. *Old Wires and New Waves.* D. Appleton-Century Company, Inc., New York, London, 1936.

Hathaway, K. A. *Television.* American Technical Society, Chicago, 1933.

Hémardinquer, Pierre. *La Télévision et Ses Progrès.* Dunod, Paris, 1933.

Hutchinson, R. W. *Television Up-to-Date.* University Tutorial Press, London, 1935.

Jenkins, Charles F. *Radiomovies, Radiovision, Television.* National Capital Press, Inc., Washington, D. C., 1929.

Jome, Hiram L. *Economics of the Radio Industry.* A. W. Shaw Co., Chicago and New York, 1925.

Langdon-Davies, John. *Radio; the story of the capture and use of radio waves.* Dodd, Mead & Co., New York, 1935.

Le Roy, Howard S. *Air Law.* Press of A. C. Melluhampe, Inc., Washington, D. C., 1936.

Lewis, E. J. G. *Television; technical terms and definitions.* Sir I. Pitman & Sons, Ltd., London, 1937.

Lodge, Sir Oliver Joseph. *Talks About Radio.* George N. Doran Co., New York, 1925.

Mata, Enrique, and Gonzales, S. F. *La Televisión*. Talleres Espasa-Calpe, Madrid, 1929.

Mesny, René. *Télévision et Transmission des Images*. A. Colm, Paris, 1933.

Millikan, Robert A. *Radio's Past and Future*. University of Chicago Press, Chicago, 1931.

Moseley, S. A., and McKay, H. *Television*. Oxford University Press, London, New York, 1936.

Moseley, S. A., and Chapple, H. J. B. *Television Today and Tomorrow*. Sir I. Pitman & Sons, Ltd., London, 1934.

Myers, Leonard M. *Television Optics*. Sir I. Pitman & Sons, Ltd., London, 1936.

National Advisory Council on Radio in Education. Advisory Committee on Engineering Developments. University of Chicago Press, Chicago, 1936.

National Committee on Education by Radio. *Proceedings of a National Conference on the Use of Radio as a Cultural Agency in a Democracy*. Edited by Tracy F. Tyler. The National Committee on Education by Radio, Washington, D. C., 1934.

Osborn, E. G. *Television for You*. Practical Press, Ltd., London, 1935.

Page, Arthur W., Arnold, H. D. and others in the Bell System. *Modern Communication*. Houghton, Mifflin Co., New York and Boston, 1932.

Radio Corporation of America. *The Radio Decade*. Radio Corporation of America, New York, 1930.

RCA Institutes, Inc. *Television*. R.C.A. Institutes Technical Press, New York, 1936.

Radio Industry. Series of Lectures given at Graduate School of Business Administration, Harvard University. Introduction by David Sarnoff. A. W. Shaw Co., Chicago and New York, 1928.

Reu, Agustín. *Radio Ciencia*. Editorial Radio, Barcelona, 1932.

Reyner, J. H. *Television, Theory and Practice*. Chapman & Hall, Ltd., London, 1934.

Richards, Vyvyan. *From Crystal to Television, the Electron Bridge*. A. & C. Black, Ltd., London, 1928.

Robinson, Ernest H. *Televiewing*. Selwyn Blount, Ltd., London, 1935.

Sarnoff, David. *Relationship of Radio to the Problem of National Defense*. Reprinted from the *United States Daily*, Washington, D. C., March 21, 1927.

Scroggie, M. G. *Television*. Blackie & Son, Ltd., London and Glasgow, 1935.

Sheldon, H. H., and Grisewood, E. N. *Television; present methods of picture transmission*. D. Van Nostrand Co., Inc., New York, 1929.

Tiltman, R. F. *Television for the Home*. Hutchinson & Co., Ltd., London, 1927.

Tiltman, R. F. *Baird of Television; the life story of John Logie Baird*. Seeley, Service & Co., Ltd., London, 1933.

Williams, Archibald. *Telegraphy and Telephony*. T. Neilson & Sons, Ltd., London, New York, etc., 1928.

Yates, Raymond F. *ABC of Television*. The Norman W. Henley Publishing Co., New York, 1929.

Notes

CHAPTER I

1. National Resources Committee, *Report on Technological Trends and National Policy*, pp. 32-33.
2. *New York Times*, Jan. 24, 1937.
3. National Resources Committee, op. cit., p. 229.
4. *Radio Today*, January, 1938.
5. N. W. Ayer & Sons, *Directory of Newspapers and Periodicals*.

CHAPTER II

1. *New York Times*, December 19, 1937.
2. *Report of the Television Committee. Presented by the Postmaster-General to Parliament by Command of His Majesty*, January, 1935.
3. *Variety Radio Directory*, 1937–1938, p. 778.
4. *New York Times*, January 24, 1937.
5. *London Times*, March 18, 1937.
6. *New York Times*, March 21, 1937.
7. *Discovery* (British), May, 1933.
8. Wolf Franck, "Rundfunkimperialismus." *Neue Weltbuehne*, October 28, 1937.
9. *Business Week*, June 8, 1935.
10. Wolf Franck, op. cit.
11. *Variety Radio Directory*, 1937–1938, p. 778.
12. *Broadcasting*, April 15, 1937.
13. *Radio News*, March, 1937.

14. *The Trans-Pacific*, Tokyo, May 19, 1932.
15. *New York Times*, March 7, 1937.
16. *Variety Radio Directory*, 1937–1938, p. 778.
17. Wolf Franck, op. cit.
18. *New York Times*, March 21 and 28, 1937.
19. H. R. 4281, 75th Congress, 1st Session.

CHAPTER IV

1. Federal Communications Commission (F.C.C.), *Informal Engineering Conference*, Docket No. 3929, Vol. I, p. 43.
2. U. S. Senate Committee on Interstate Commerce, *Hearings on S. 6*. 71st Congress, 1st Session, p. 92.
3. F.C.C., op. cit., pp. 71-72.
4. Ibid., p. 44.

CHAPTER VI

1. U. S. Senate Committee on Interstate Commerce, *Hearings on S. 6*. 71st Congress, 1st Session, p. 86.
2. U. S. Court of Appeals for the District of Columbia 6852, 3, 4; Great Western Broadcasting Assoc. Inc. vs. Federal Communications Commission, et al. 1937.
3. U. S. Senate Committee on Interstate Commerce, op. cit., p. 56.
4. Public Law No. 262; 61st Congress, June 24, 1910.
5. U. S. Senate Committee on Interstate Commerce, op. cit.
6. Public Law No. 264; 62nd Congress, August 13, 1912.
7. U. S. Senate Committee on Interstate Commerce, op. cit., pp. 62-63.
8. Ibid., pp. 63-64.
9. Ibid., p. 64.
10. Ibid., pp. 64–65.
11. Ibid., pp. 65-66. (The Tribune Co. vs. Oak Leaves Broadcasting Station Inc., et al.)
12. Ibid., p. 66.

NOTES 285

13. Public Law No. 632; 69th Congress, February 23, 1927 (Federal Radio Act).
14. Public Law No. 416; 73rd Congress, June 19, 1934 (Communications Act).

CHAPTER VII

1. F.C.C., *Informal Engineering Conference.* Docket No. 3929, Vol. V, pp. 876-7, June 19, 1936.
2. Ibid., Vol. III, p. 369, June 17, 1936.
3. F.C.C., *Annual Report for 1937*, p. 185.
4. Ibid.
5. F.C.C.; In the matter of: Petition of Purdue University, W9XG, July 14, 1936.

CHAPTER VIII

1. F.C.C., Press Release No. 23463, October 13, 1937.
2. F.C.C., *Informal Engineering Conference*, Vol. II, pp. 252-3, June 16, 1936.
3. Ibid., p. 163.
4. *Congressional Record*, July 19, 1937.
5. F.C.C., *Informal Engineering Conference*, Vol. III, p. 486, June 17, 1936.
6. Ibid., Vol. IV, pp. 709-13, June 18, 1936.
7. Ibid., Vol. III, pp. 530-535, June 17, 1936.
8. *New York Times*, April 18, 1937.
9. F.C.C., op. cit., Vol. I, pp. 23-32, June 15, 1936.

CHAPTER IX

1. *Broadcasting Year Book*, 1937, p. 231.
2. *Variety Radio Directory*, 1937–1938, p. 682.
3. F.C.C., *Hearings Pursuant to Sec. 307 of the Communications Act*, October 1-20, 1934, and November 7-12, 1934.
4. F.C.C., Release No. 11861, letter E. O. Sykes to President of the Senate, January 22, 1935.

5. Brief of the City of New York (WNYC) in the Court of Appeals, Dist. of Columbia, filed December 13, 1932; City of New York vs. Federal Radio Commission, et al.
6. See 4 above.
7. Ibid.
8. F.C.C., *Annual Report for 1937*.
9. *Variety Radio Directory, 1937-1938*, p. 785.
10. F.C.C., *Informal Engineering Conference*, Vol. I, pp. 105-126, June 15, 1936.
11. House Committee on Appropriations, *Hearings on Independent Offices Appropriations Bill for 1939*; 75th Congress, 2nd Session, p. 1240.
12. *Literary Digest*, July 4, 1936.
13. *Broadcasting Year Book*, 1937, p. 123.
14. U. S. Dept. of Commerce, Census of Business 1935, *Report on Radio Broadcasting*.
15. House Committee on Appropriations, *Hearings on Independent Offices Appropriations Bill for 1938*; 75th Congress, 1st Session, p. 376.
16. Ibid.
17. *Broadcasting Year Book*, 1937, p. 24.
18. *Congressional Record*, July 19, 1937.

CHAPTER X

1. *Variety Radio Directory, 1937-1938*, pp. 684-685.
2. Ibid.
3. Ibid.
4. Ibid.
5. Ibid.
6. Ibid.
7. *London Times*, March 17, 1936.
8. David Sarnoff, Speech before the Conference on American Foreign Policy, Cambridge, Mass., December 3, 1937.
9. American Civil Liberties Union, *Radio Is Censored!*
10. Ibid.

11. Ibid.
12. Ibid.
13. Ibid.
14. Ibid.
15. Ibid.
16. Ibid.
17. Ibid.
18. Ibid.
19. Ibid.
20. Ibid.

CHAPTER XI

1. *Broadcasting Year Book*, 1937.
2. Ibid.
3. Ibid.
4. George Henry Payne, Address before Second National Conference on Education, Chicago, Ill., December 1, 1937.
5. *Variety*, December 7, 1937.

CHAPTER XII

1. *Broadcasting*, June 15, 1937.
2. Academy of Motion Picture Arts and Sciences, *Report on Television*, 1936.
3. F.C.C., Special Telephone Investigation Docket No. 1 Exhibit No. 304. Contract between Electrical Research Products, Inc. and various motion picture producers.
4. Ibid. Exhibit 504, Letter from J. E. Otterson, President of Electrical Research Products, Inc. to E. S. Bloom, President of Western Electric Co., November 5, 1932.
5. Ibid. Exhibit 296, Letter Otterson to Bloom, December 7, 1933.
6. **Ibid.** Exhibit 306, Letter Otterson to Bloom, April 29, 1937.

CHAPTER XIII

1. F.C.C., Special Telephone Investigation Docket No. 1. Report on American Telephone and Telegraph Company, Corporate and Financial History, p. 51.
2. F.C.C., Spec. Tel. Inv. Report on Scope and Structure of the Bell System, pp. 9-21.
3. Ibid.
4. F.C.C., Spec. Tel. Inv. Report on Patent Structure of Bell System.
5. F.C.C., Spec. Tel. Inv. Report on American Telephone Company, Transoceanic Radio Telephony. Exhibit 2093, p. 45.
6. New York Times, December 5, 1934.
7. F.C.C., Spec. Tel. Inv. Exhibit 578, "Four Square Memorandum" by J. E. Otterson, dated January 13, 1927.
8. Alvin Harlow, Old Wires and New Waves, pp. 356-358.

CHAPTER XIV

1. F.C.C., Spec. Tel. Inv. Report on American Telephone and Telegraph Company, Corporate and Financial History, p. 11.
2. Ibid., pp. 11-12.
3. Ibid., Chap. II.
4. Ibid., p. 12.
5. Alvin Harlow, Old Wires and New Waves, pp. 377-379.
6. F.C.C., op. cit., p. 205.
7. Ibid., p. 21.
8. Ibid., Appendix 7. Agreement of November 10, 1879, between the Western Union Telegraph Co. and the National Bell Telephone Co.
9. Ibid., p. 26.
10. F.C.C., Spec. Tel. Inv. Report on American Telephone and Telegraph Company, Origin and Development of License Contract.

11. F.C.C., Spec. Tel. Inv. Report on Control of Telephone Communications, Vol. I, p. 65.
12. Ibid., Vol. V, p. 101.
13. Alvin Harlow, op. cit., p. 455.
14. F.C.C., Spec. Tel. Inv. Report on A. T. & T. Co. Corporate and Financial History, pp. 255-257.
15. F.C.C., Spec. Tel. Inv. Report on Scope and Structure of the Bell System, p. 24.

CHAPTER XV

1. House Committee on Patents, Hearings on H. R. 4523, 74th Congress, Vol. II, p. 1340.
2. Ibid., p. 1349.
3. F.C.C., Spec. Tel. Inv. Report on Electrical Research Products, Inc., Chap. VIII.
4. Ibid., p. 279.
5. Ibid., p. 293.
6. F.C.C., Spec. Tel. Inv. Exhibit 379.
7. F.C.C., Spec. Tel. Inv. Report on Electrical Research Products, Inc., p. 361.
8. Ibid., pp. 475-478.
9. Ibid., pp. 472-473.
10. Ibid.
11. 249 U. S. 464, 294 U. S. 464 (1934).
12. F.C.C., op. cit., Chap. X.
13. Ibid.
14. Ibid., p. 249.

CHAPTER XVI

1. U. S. Senate Committee on Interstate Commerce, Hearings on S. 6, 71st Congress, 1st Session, p. 1108.
2. Ibid., p. 1108.
3. Ibid., p. 1109.

290 NOTES

4. Ibid., p. 1358-1359.
5. London Times (see London Times Index for 1918 under Marconi Co.).
6. Ibid., p. 314.
7. Moody's Manual of Investments 1937, Radio Corp. of America, p. 2187-2191.
8. U. S. Senate Committee on Banking and Currency, Hearings on Stock Exchange Practices, 72nd Congress, 1932-1933, pp. 468-9.
9. F.C.C., Spec. Tel. Inv. Exhibit 289 A6.
10. F.C.C., Spec. Tel. Inv. Report on Bell System's Policies and Practices in Radio Broadcasting, p. 84.
11. Ibid., pp. 88-89.
12. Ibid., pp. 24-32.
13. Op. cit., Exhibit 289 A 19.

CHAPTER XVII

1. F.C.C., Spec. Tel. Inv. Report on Bell System's Policies and Practices in Radio Broadcasting, pp. 32-55.
2. Petition In Equity No. 793, U. S. Dist. Court, Del. United States of America vs. Radio Corporation of America, et al. (1930).
3. F.C.C., op. cit., pp. 2-3.
4. Federal Trade Commission, Report on the Radio Industry. 1923.
5. U. S. Senate Committee on Interstate Commerce, Hearings on S. 6, 71st Congress, p. 1196.
6. Public Law No. 632, 69th Congress, Sec. 13. February 23, 1927.
7. S. 6. 71st Congress.
8. Lord vs. RCA. 24 F (2d) 565 affirmed 28 F (2d) 257.
9. Federal Radio Commission, Release No. 4910, June 24, 1931.
10. F.C.C., Spec. Tel. Inv. Exhibit 577.
11. Ibid., Exhibit 1429.

12. F.C.C., Spec. Tel. Inv. *Report on Electrical Research Products, Inc.*, p. 131.
13. F.C.C., Spec. Tel. Inv. Exhibit 289 A 35.

CHAPTER XVIII

1. House Committee on Patents, *Hearings on H. R. 4523*, 74th Congress, p. 564.
2. Ibid., Testimony of Waldemar Kampffert, p. 895.
3. Floyd L. Vaughn, *Economics of Our Patent System*, p. 181.
4. Pennock vs. Dialogue, 2. Pet. 1.
5. 167 U. S. 224.
6. F.C.C., *Informal Engineering Conference*, Vol. II, p. 218, June 16, 1936.
7. Westinghouse Electric and Manufacturing Co. vs. RCA; In Equity No. 1183, U. S. Dist. Court of Del., July 9, 1936.
8. U. S. Patent Office; Patent Interference No. 64207, P. T. Farnsworth vs. V. Zworykin.
9. Ibid.
10. Ibid.
11. U. S. Patent Office; Patent Interference No. 62721, Round vs. Zworykin vs. Jenkins vs. Mathes.

CHAPTER XIX

1. Rules and Orders of F.C.C.
2. "Philco." *Fortune*, February, 1935, p. 75.
3. Complaint of Philco Radio and Television Corp. filed in Supreme Court of the State of N. Y. County of New York; *Philco Radio and Television Corp. vs. Radio Corporation of America, et al.*, July 30, 1936.
4. *Business Week*, April 17, 1937.
5. F.C.C., Spec. Tel. Inv. *Report on Bell System Policies and Practices in Radio Broadcasting.*

292 NOTES

CHAPTER XX

1. F.C.C., *Informal Engineering Conference*, Vol. I, pp. 122-3, June 15, 1936.
2. Ibid., Vol. II, pp. 174-175, June 16, 1937.
3. Ibid.
4. Ibid.
5. *Literary Digest*, July 4, 1936.
6. T. R. Carskadon, "Report on Television." *Theatre Arts Monthly*, June, 1937.
7. F.C.C., op. cit., Vol. VI, pp. 1183-1184, June 21, 1936.

CHAPTER XXI

1. U. S. Senate Committee on Interstate Commerce, *Hearings on Confirmation of Members of the Federal Communications Commission*, 74th Congress, 1st Session, p. 8.
2. Ibid., p. 167.
3. Ibid., p. 84.
4. F.C.C., *Report on Social and Economic Data Pursuant to the Informal Hearing on Broadcasting*. Docket No. 4063, July 1, 1937.
5. F.C.C., In Re: Application of Mackay Radio and Telegraph Co., Inc. Docket Nos. 3336, 3337, 3338.

CHAPTER XXII

1. *Vital Speeches*, July 15, 1936, p. 664.

Index

Academy of Motion Picture Arts and Sciences, 126, 129
Advertising, cost, 265
Alexanderson, Dr. E. F. W., 10, 162; alternator, 165
Alexandra Palace, 12, 13
Algeria, 98
Alien Property Custodian, 166
Aloisi, Baron, 165
American Bar Assoc., 49, 55
American Bell Telephone Co., 144
American Civil Liberties Union, 103, 108
American Council on Education, 5
American Legion, 105
American Marconi Company, 52, 163, 164, 166
American Medical Assoc., 74
American Speaking Telephone Co., 143
"American System," 252, 265
American Telephone & Telegraph Co., 9, 28, 42, 209; activities in motion picture industry, 126-130, 150-161; agreement with Farnsworth, 213; arbitration with RCA, 175; coaxial cable, 198, 217; cross-licensing agreement of 1920, 170, of 1926, 178; F.C.C. investigation, 247; government anti-trust suit, 179, 196; history, 139-148; liquidation of RCA, 130; ownership and sale of RCA stock, 178; plan for control of radio, 172; scope, 131; trading philosophy, 135. See also Bell system
Armistice Day Incident, 12
Armstrong, 10
Associated Press, 52
Audion tube, 27
Australia, 98

Baird, J. L., 10
Baker, George F., 146
Baker, Newton D., 223
Baldwin, James W., 233
Beaverbrook, Lord, 134
Bell, Alexander Graham, 137, 139
Bell Patent Assoc., 140
Bell system, 10, 56, 131-61. See also American Telephone & Telegraph Co.
Bell Telephone Assoc., 141
Bell Telephone Laboratories, Inc., 28, 42, 150
Bellows, Henry Adams, 104
Berlin, treaty of, 52
Berliner, Emil, 142
Bergen, Edgar, 94
Berne, treaty of, 65
Bilbo, Senator T. G., 244
Black, Hugo, 112
Bliss, Cornelius, 223

294   INDEX

Bloom, E. S., 129, 134
Bowes, Major Edward, 95
Boyden, Roland W., 175
Braun, Arthur E., 223
Brazil, 97
Brigham Young University, 212
British Broadcasting Corp., 12, 97, 99, 101
British Marconi Co., 163, 164
Browder, Earl, 106
Brown Bros., 224
Brown, Thad H., 88, 243-5
Bullard, Admiral W. H. G., 166
Bush, Prescott S., 224
Business Week, 220
Butler, Smedley D., 109

Cadman, S. Parkes, 90
Cairo Conference, 45
Caldwell, Louis G., 49
Canada, 83
Canadian Broadcasting Corp., 98
Carter, Boake, 105
Case, Norman C., 248-9
Censorship, 97-110
Chase National Bank, 128
Chicago Federation of Labor, 92
Chile, 97
China, 98
Columbia Broadcasting System, 9, 72; agreement with Farnsworth, 213; banking connections, 224; code of ethics, 115; international broadcast licenses, 92; M.P.P.D.A. plan, 125; percentage of power, 96; programs, 220; stock issue, 73
Conrad, Dr. Frank, 171, 180
Coolidge, Calvin, 108
Coronation, 12
Cost of television sets, 12
Coughlin, Father C. E., 110
Cravath, Paul D., 153
Craven, Tunis A. M., 249-52

Crosley Radio Corp., 92
Cross-licensing agreement of 1920, 169; of 1926, 178; of 1932, 192, 197
Cuba, 83
Cutler, Bertram, 223
Czechoslovakia, 15

Darby, Samuel E., 209, 210
Davis, Col. Manton, 181, 182
Davis Cup matches, 12
De Forest, Lee, 10, 26, 38, 185, 216; on programs, 116; television developments, 239-40
De Forest Radio Co., 185
Delano, Frederic A., 4
Dellinger, J. H., 79
Dolbear, E. A., 140
Dominion Theatre, 13

Edison effect, 25
Edison, Thomas Alva, 24, 37, 206
Educational stations, see Appendix B
Electrical Research Products, Inc., 128, 136, 145, 150, 186; production of pictures, 160
Electronics Foundation, 195
Elizabeth, Queen, 201
Encyclopedia of the Social Sciences, 17
England, 12
Espenschied, Lloyd, 77-9
Everson, 212

Facsimile, 8, 17, 238
Farnsworth, Philo, 10, 199, 217; agreement with A. T. & T. Co., Philco, and Columbia, 213; ally of Bell laboratories, 198; connection with television in England, Germany and the U. S., 208; image dissector, 32; interference action with Zworykin, 210

INDEX 295

Farnsworth Television Co., 238
Federal Communications Act, 57-61
Federal Communications Commission, conference on non-commercial broadcasting of 1934, 86-90; description of members, 241-53; special telephone investigation, 247
Federal Radio Commission, 56, 85; Lord case, 188
Federal Radio Education Committee, 90
Federal Trade Commission, 180, 183
Federalist, 201
Fernsee, 209
Fessenden, R. A., 147, 149
Finch, William, 10
Fish, Frederick P., 145
Fish, Hamilton, 106
Fleming, J. A., 25, 37
Folk, George E., 194, 195
Fox, William, 123, 150; loan of $15,000,000 by A. T. & T. Co., 157; purchase of Tri-Ergon patents, 158
Fox-Case Corporation, 151
France, 14; station in Eiffel Tower, 13
Franklin, Benjamin, 201

General Electric Co., 135, 162; attack in Congress, 224; Congressional attack, 229; consent decree, 196; cross-licensing agreement of 1920, 169; government anti-trust suit, 179; international broadcast licenses, 92; ownership of RCA stock, 178; sale of Alexanderson alternator, 165
General Motors Radio Corp., 179
Germany, 14; Communist radio station, 19; plan to put radio and television on wires, 19

Gifford, Walter S., 154
Gold and Stock Co., 141
Gorrell, 212
Gould, Jay, 140
Government Ownership, 268
Graves, V. Ford, 74
Gray, Elisha, 137, 139
Green, William, 90, 238
Griffin, Admiral, 162
Griswold, A. H., 172, 173

Hahne, C. A., 218
Halsey, Stuart & Co., 159
Harbord, James G., 166
Hard, William, 105
Harden, Edwin, 224
Harding, C. F., 67
Harley, John S., 218
Harriman & Co., 224
Harvard University, 92
Hays, Will, 124, 152
Heintz, Ralph M., 64
Holland, 15
Hollywood Reporter, 228
Hopkins, Harry L., 4
Hubbard, Gardiner C., 139, 140, 141, 142
Hughes, Charles E., 159
Hutton & Co., 224

Ickes, Harold L., 4
Iglehart, Joseph A. M., 224
International Association of Police Chiefs, 72
International broadcasting licenses, 92
International Bureau of Standards, 79
International Council of Religious Education, 87
International Telegraph & Telephone Co., 135
Ipswich, 13

296 INDEX

Isle of Dreams Broadcasting Corp., 92
Italy, 15
Izaac Walton Club, 108

James I, 201
Japan, 15
Jefferson, Thomas, 203
Jewett, F. B., 28, 41, 42, 44, 193, 196, 216
Johnson, Father G. W., 90
Johnson, Hugh, 107
Jolliffe, C. B., 75-7
Jolson, Al, 150

Kaiser Wilhelm of Germany, 51
Kennelly-Heaviside layer, 23, 63
Kent, Sidney, 152
Kestler, Laurence, Jr., 218
KMMJ, 81
KNX, 94, 244
Ku Klux Klan, 112

Lafount, Commissioner, 188
La Mert, R. D., 10, 240
Langley, Professor, 205
Law; of 1910, 52; of 1912, 53; 1923 test of, 53; Communications Act of 1934, 57-61; inadequacy of, 49-61; patent law of 1790, 203, of 1836, 204, of 1870, 204; Radio Act of 1927, 56
Lawrence, David, 106
Leavy, Rep. C. H., 227
Lehman Bros., 224, 244
Leonard, Capt. Donald S., 72
Llewellyn, George, 244
"Lock and key," 231
Loew's, Inc., 157
Lord, Arthur, D., 186

Mackay Radio & Cable Co., 10; Oslo case, 253; rate hearing, 252
Madison, James, 201

Marconi Company of America, see American Marconi Co.
Marconi, Gugliemo, 38, 147
Marine use, 9
Mayer, Louis B., 152
McFarlane, Rep. W. D., 222-30
McNinch, Frank R., chairman of F.C.C., 242; Mae West incident, 258
Meller, Raquel, 150
Merriam, Charles E., 4
Metro-Goldwyn-Mayer, 157
Mexico, 83
Michelson, Charles, 84
Military use, 9
Millhauser, Dewitt, 224
Millikan, R. A., 90
Morgan, J. P. & Co., 223
Morgan-Baker Group, 146
Morocco, 98
Motion Picture Industry, 121-30
Motion Picture Producers & Distributers of America, 124
Mouromtseff, Mr., 213
Mutual Broadcasting System, 96, 125, 222
Myers, W. E., 105

National Academy of Sciences, 5
National Association of Broadcasters, 87; code of ethics, 114; on regulation of television, 233-5
National Association of State Universities, 87
National Bell Telephone Co., 143
National Broadcasting Co., 7; code of ethics, 115; consent decree, 196; creation of, 178; international broadcast licenses, 92; M.P.P.D.A. plan, 125; Mae West incident, 118, 258; percentage of power, 96
National Catholic Educational Assoc., 87

INDEX 297

National Conference on Educational Broadcasting, 115
National Educational Assoc., 87
National Recovery Act, 201
National Resources Committee, 5
Naval television, 9
New England Telephone Co., 142
Newspapers, competition with radio, 7, 8; ownership of radio stations, 227
N. Y. Philharmonic Society, 220
Nields, John P., 192, 195
Nipkow, 22, 30

Oklahoma State Corporation Commission, 247
Olympic Games; 1936 Germany, 14; 1940 preparation by Japan, 15
Otterson, J. E., 129, 134, 136; Fox trustee, 159; president of Paramount Pictures, 156; relations with RCA and ERPI, 153; with Warner Bros., 151; Tri-Ergon patents, 158

Paley, William S., 72, 73, 220, 265; banking interest in Columbia, 224
Paramount Picture Co., 224
Parran, Dr. Thomas, 107
Patent pools, 12
Patents, 199-215; cost of litigation, 225; law of 1790, 203, of 1836, 204, of 1870, 204
Patman, Rep. Wright, 223
Payne, George Henry, 115, 246
Peabody, Eddie, 150
Peoples Telephone Company of N. Y., 144
Perkins, Frances, 4
Philadelphia Storage Battery Co., 217, 218
Philco Radio & Television Corp., 209, 213, 217, 222; suit against RCA, 218

Poland, 15
Postal Telegraph & Cable Co., 10, 174; rate hearing, 252; revenue from radio, 178
Prall, A. Mortimer, 124, 125, 217
Prall, Anning, 249
Purdue University, 67

Radio Corporation of America, 9, 39, 75, 209, 222; activities in motion picture industry, 152-6; attack in Congress, 223-5; competition with A. T. & T. Co., 161; consent decree, 196; cross-licensing agreement of 1920, 171, of 1926, 178; government antitrust suit, 179; history, policies, scope, 162-98; joined by Zworykin, 214; Lord case, 187; order for equipment from Russia, 15; photophone, 152; protection against by A. T. & T. Co., 135; royalties from independents, 200; stock transfer to G. E. and Westinghouse, 229; suit against A. T. & T. Co., 154; suit by Philco, 218; suit by Westinghouse, 215
Radio Manufacturers Assoc., 64; on regulation of television, 235-7
"Radio Trust," 178-98, 224; attack in Congress, 224; price of sets, 225
Radio-Keith-Orpheum, 152
Radiolympia, 12
RCA Communications Co., 252, 253
RCA Photophone, 179
RCA-Victor Co., 179
Regulation of television, 259-69
Reynolds, Walter L., 104
Rich, Walter J., 150
Richardson, Dorsay, 224
Robertson, A. M., 223
Robins, Robert, 267

Robinson, Commissioner, 188
Rockefeller, John D., 224
Roosevelt, F. D., 112, 162, 164, 166, 167
Roper, Daniel C., 4
Round, Henry Joseph, 10, 210, 214
Ruml, Beardsley, 4
Russia, 14, 15
Russian Society of Wireless Telephone & Telegraph, 213

SAFAR, 15
Sanders, Thomas, 139, 141
Sarnoff, David, 44, 191, 196, 216; American way, 265; criticizes Jewett, 42; on foreign broadcasting, 102; on the future of radio, 39; president of RCA, 167; suit against A. T. & T. Co., 154; on television for message service, 134; *Titanic* disaster, 52
Schuette, Oswald, 179, 180, 182, 186, 196
Screen, size of, 13
Securities & Exchange Commission, 73, 229
Segal, Paul M., 246
Seldes, Gilbert, 220
Seligman, J. W. & Co., 224
Shaw, Fred B., 108
Sheffield, J. R., 224
Skinner, James M., 235-7
Social Science Research Council, 5
Sokolsky, George, 105
Spectrum, 62-9
Speyer & Co., 224
Standards, British, 12, 13
Starbuck, Commissioner, 188
Stations, television, see Appendix A
Statistics, radio, 93
Strauss, Frederick, 224
Stuart, Harry, 159
Sweden, 15

Sykes, Eugene O., member of F.C.C., 88, 190, 243-5

Tax on Radio, 226
Taxation, British, 13
Taylor, M. Sayle, 95
Television stations, see Appendix A
Thayer, H. B., 146
Thomas, Norman, 110
Thompson, Dorothy, 105
Thompson, J. J., 25
Titanic disaster, 52
Tri-Ergon patents, 158

Ullswater, Lord, 99
U. S. Dept. of Justice, 155, 178, 183; suit against "radio trust," 179
U. S. Dept. of Labor, 87
U. S. Interior Dept., Bureau of Education, 20
U. S. Navy Dept., 74, 163, 164, 166
U. S. Patent Office, 137, 204, 205, 210
U. S. Senate Committee on Interstate Commerce, 183
U. S. State Dept., 20
Upton Sinclair Presents William Fox, 151

Vail, Theodore N., 142, 143, 146; president of Western Union, 147
Vandenberg, Senator, 107
Vanderbilt, William H., 140
Vanderlip, Frank, 224
Variety, 116
Versailles Treaty, 5
Vitaphone Corp, 127, 150
Voorhis, Rep. J., 226

Walker, Paul A., 247
Wallace, Henry A., 4
Warner Bros., 150, 151

Washington, George, 203
Waterbury, J. I., 146
Watson, Thomas A., 139
WCAU, 92
WEAF, 173, 196
Wearin, Rep., 227
Weeks and Hardin, 224
West, Mae, 118, 258
Western Electric Co., 150, 151, recording equipment, 152
Western Union Telegraph Co., 10, 135; early relations with Bell System, 140; 1879 contract with Bell system, 143; control by A. T. & T. Co., 147; radio wires, 174; rate hearing, 252; revenue from radio, 178
Westinghouse Electric & Mfg. Co., 9, 171, 209, 214; Congressional attack, 229; government anti-trust suit, 179; international broadcast licenses, 92; ownership of RCA stock, 178; patent suit against RCA, 215
Wheeler, Senator Burton K., 245
WHN, 106
Wigglesworth, Rep., 226
Wilson, Woodrow, 162
Winchell, Walter, 109
WIRE, 106
WJAZ, 54
WLW, 84
WMCA, 89
WNYC, 88
Woodring, Harry H., 4
Worldwide Broadcasting Corp., 92

Yankee Network, 106
Young, Owen D., 164, 184, 196; on freedom of speech, 103

Zenith Radio Corp., 54
Zworykin, Vladimir, 10, 34, 199, 208, 210, 213

HISTORY OF BROADCASTING:
Radio To Television
An Arno Press/New York Times Collection

Archer, Gleason L.
Big Business and Radio. 1939.

Archer, Gleason L.
History of Radio to 1926. 1938.

Arnheim, Rudolf.
Radio. 1936.

Blacklisting: Two Key Documents. 1952–1956.

Cantril, Hadley and Gordon W. Allport.
The Psychology of Radio. 1935.

Codel, Martin, editor.
Radio and Its Future. 1930.

Cooper, Isabella M.
Bibliography on Educational Broadcasting. 1942.

Dinsdale, Alfred.
First Principles of Television. 1932.

Dunlap, Orrin E., Jr.
Marconi: The Man and His Wireless. 1938.

Dunlap, Orrin E., Jr.
The Outlook for Television. 1932.

Fahie, J. J.
A History of Wireless Telegraphy. 1901.

Federal Communications Commission.
Annual Reports of the Federal Communications Commission.
1934/1935–1955.

Federal Radio Commission.
Annual Reports of the Federal Radio Commission. 1927–1933.

Frost, S. E., Jr.
Education's Own Stations. 1937.

Grandin, Thomas.
The Political Use of the Radio. 1939.

Harlow, Alvin.
Old Wires and New Waves. 1936.

Hettinger, Herman S.
A Decade of Radio Advertising. 1933.

Huth, Arno.
Radio Today: The Present State of Broadcasting. 1942.

Jome, Hiram L.
Economics of the Radio Industry. 1925.

Lazarsfeld, Paul F.
Radio and the Printed Page. 1940.

Lumley, Frederick H.
Measurement in Radio. 1934.

Maclaurin, W. Rupert.
Invention and Innovation in the Radio Industry. 1949.

Radio: Selected A.A.P.S.S. Surveys. 1929–1941.

Rose, Cornelia B., Jr.
National Policy for Radio Broadcasting. 1940.

Rothafel, Samuel L. and Raymond Francis Yates.
Broadcasting: Its New Day. 1925.

Schubert, Paul.
The Electric Word: The Rise of Radio. 1928.

Studies in the Control of Radio: Nos. 1–6. 1940–1948.

Summers, Harrison B., editor.
Radio Censorship. 1939.

Summers, Harrison B., editor.
A Thirty-Year History of Programs Carried on National Radio Networks in the United States, 1926–1956. 1958.

Waldrop, Frank C. and Joseph Borkin.
Television: A Struggle for Power. 1938.

White, Llewellyn.
The American Radio. 1947.

World Broadcast Advertising: Four Reports. 1930–1932.